Extending the Life of Reservoirs

DIRECTIONS IN DEVELOPMENT
Energy and Mining

Extending the Life of Reservoirs

Sustainable Sediment Management for Dams and Run-of-River Hydropower

George W. Annandale, Gregory L. Morris, and Pravin Karki

WORLD BANK GROUP

© 2016 International Bank for Reconstruction and Development / The World Bank
1818 H Street NW, Washington, DC 20433
Telephone: 202-473-1000; Internet: www.worldbank.org

ISBN (paper): 978-1-4648-0838-8
ISBN (electronic): 978-1-4648-0837-1
DOI: 10.1596/978-1-4648-0838-8

Cover photo: © Gregory L. Morris. Used with permission. Further permission required for reuse.
Cover design: Debra Naylor, Naylor Design, Inc.

Library of Congress Cataloging-in-Publication Data has been requested

Contents

Boxes

Figures

Maps

Photos

Tables

Foreword

Economic development, now and in the future, relies critically on infrastructure development. For example, if well planned, hydropower facilities and dams provide water supply, irrigation capacity, and renewable sources of electricity. Yet without careful planning, including management of river basin sediments, the services provided by hydropower facilities and dams are at risk. Ensuring the long-term resilience of these critical infrastructure facilities requires early and consistent attention to the processes of reservoir sedimentation, which reduce the storage capacity of reservoirs and damage hydromechanical equipment, posing a threat to the sustainability of hydropower, water supply, and irrigation services.

This book is welcome because it provides guidance on adopting sustainable sediment management practices for hydropower and water supply dam projects. Fortunately, effective sediment management techniques, particularly those adopted as part of hydropower project design, can cost-effectively counteract these effects.

The focus audience of this document is policy makers, lending agencies, and general practitioners. The level of detail provided should appeal to all, as it falls somewhere between the extensive and exhaustive material already available in the scientific literature and the often very simplistic summary reports that fall short of providing practical guidance.

This book gives people working on dams an argument as to why it is so important to think of sediment when we support planning and implementation of dams, and shows why sediment is an issue to consider very early in the decision-making and design process. It provides a means by which to check the solution suggested by a developer and compare it with other methods. It is not easy for laypeople to judge whether the developer's stated solution actually is feasible. This book helps the development practitioner to be better informed in evaluating dam and hydropower proposals.

While this report—written by two of the world's leading technical experts on the subject who have been involved in many large projects financed by the World Bank—is an excellent introduction to the technical aspects of sediment management, it also adds a new perspective not found in previous work: the joint effects of climate change and storage loss due to reservoir sedimentation. It is useful to understand that sediment management measures are a robust adaptation strategy for supporting sustainable hydropower, water supply, and irrigation services. These measures make sense regardless of future climate, but in many cases have even more value when uncertainty about future hydrological patterns is taken into account.

Anita George
Senior Director (former)
Energy and Extractives Global Practice
World Bank

Acknowledgements

Publication of this book would not have been possible without contributions from two of the world's leading experts in sediment management, George W. Annandale and Gregory L. Morris. The book is essentially a distillation of their lifelong work in the field.

Pravin Karki, coordinator of this book project, is grateful to Julia Bucknall, Director of the World Bank's Environment and Natural Resources General Practice for providing guidance and constant support. The team is grateful to Marianne Fay, Chief Economist, Sustainable Development Practice Group for chairing the decision meeting and providing guidance on discount rates. The team thanks peer reviewers Erick Fernandes, Abedalrazq Khalil, Rikard Liden, Peter Meier, and Satoru Ueda.

We appreciate the advice and sharing of experiences with our clients from all over the world who face challenges in dealing with sediments in their daily lives. Special thanks to Jim Neumann and Margaret Black of Industrial Economics, Incorporated (IEC) for providing editorial and intellectual support. We are grateful to Mark Ingebretsen, Jewel McFadden, and other colleagues in the Publishing and Knowledge unit for their constant support in publication and dissemination work.

This publication was funded by the Energy Sector Management Assistance Program (ESMAP), the South Asia Water Initiative (SAWI), and the Asia Sustainable and Alternative Energy Program (ASTAE). ESMAP is a global multi-donor technical assistance trust fund administered by the World Bank. It provides analytical and advisory services to low- and middle- income countries to increase their know-how and institutional capacity to achieve environmentally sustainable energy solutions for poverty reduction and economic growth. Supported by the governments of the United Kingdom, Australia and Norway, SAWI is a multi-donor trust fund that works to improve the management of the major Himalayan river systems of South Asia for sustainable, fair and inclusive development, and climate resilience. Since 1992, ASTAE has been helping the East Asia and Pacific and South Asia regions transition to sustainable, inclusive, and low carbon green growth paths. Its work programs rest on three pillars of renewable energy, energy efficiency, and access to energy.

About the Authors

George W. Annandale has more than 40 years of experience as a civil engineer specializing in water resources engineering. He has published numerous peer-reviewed papers and is author, coauthor, and contributing author to eight books on sedimentation and scour. He is known for his expertise in reservoir sediment management, is a leading authority on scour of earth materials, and has extensive experience in fluvial hydraulics and sediment transport and assessing the impacts of climate change on water supply and hydropower projects. He consults globally and has worked on water resource engineering projects in more than 25 countries. The journal *International Water Power and Dam Construction* identified him as one of 20 engineers who globally made a significant contribution to the dam industry. Dr. Annandale is an active consultant on World Bank projects.

Gregory L. Morris has more than 40 years of experience working internationally as a consultant in the fields of hydrology, sedimentation engineering, and civil design. He is lead author of the *Reservoir Sedimentation Handbook*, and has performed a wide range of engineering work associated with water resources, sediment management, and environmental protection and remediation. He has worked and lectured in more than 30 countries and is known for his expertise in developing environmentally sustainable fluvial engineering strategies. Dr. Morris is an active consultant on World Bank projects.

Pravin Karki is a senior hydropower specialist at the World Bank Group. He has more than 25 years of experience as a water resources, hydraulics, and hydropower engineer in consulting, international policy, and academic settings. He has a special interest in sediment management and climate change in the water resources, hydropower, and dams sector and has experience living and working in both developing and developed countries. He has worked on World Bank hydropower projects in Africa; Central, East, and South Asia; and the Pacific. He is a core member of the Global Solutions Group on Hydropower and Dams at the World Bank.

Abbreviations

1D	one-dimensional
2D	two-dimensional
BOT	build-operate-transfer
DHP	Dasu Hydropower Project
ERR	economic rate of return
IPCC	Intergovernmental Panel on Climate Change
IRR	internal rate of return
LTCR	long-term capacity ratio
MAF	mean annual flow
mcm	million cubic meters
NPV	net present value
ROR	run-of-river
SSC	suspended sediment concentration
t/km^2/yr	tons per square kilometer per year

CHAPTER 1

Purpose and Application of This Book

Pravin Karki

Introduction

This book was developed to facilitate implementation of a programmatic approach using selected scientific methods for screening climate change and disaster risks, and integrating appropriate resilience measures into water, hydropower, and dam investment projects. As the World Bank Group steps up its activities in both the water and energy sectors, the risks of climate change and disasters need to be better understood and managed to ensure sustainable, resilient, and cost-effective outcomes. This increased awareness is particularly important for hydropower, water supply reservoir, and dam projects, given that climate change is projected to significantly affect water resources by changing mean annual river flows and hydrologic variability, thereby causing more extreme droughts and floods. For many countries, hydropower is now the largest source of affordable renewable energy (World Bank 2013b). This is especially true in regions like Sub-Saharan Africa, South Asia, and Southeast Asia, which are characterized by significant untapped hydropower potential and water shortages.

The World Bank Group's engagement in hydropower and water supply projects of all sizes and types requires careful planning to ensure resilience against the uncertainties of climate change and disaster risk. Even though sedimentation poses the greatest threat to the sustainability of hydropower, most guidelines on water supply and dam projects provide virtually no direction on how to deal with the sediment problem. Current guidelines tend to focus on mitigating changes in flows associated with climate change and do not address sedimentation.

Recognizing the importance of creating and maintaining reservoir storage, the World Bank previously developed the reservoir conservation (RESCON) approach (Palmieri et al. 2003) to facilitate rapid identification of technically sound and economically optimal reservoir sediment management strategies. Since then, greater understanding of reservoir sediment management technology has emerged, leading the World Bank to invest further in upgrading the RESCON approach. This book complements the upgraded RESCON model by providing guidance on adopting sustainable sediment management practices for hydropower and water supply dam projects.

The World Bank's Role in Sustainable Infrastructure Activities

The World Bank is intimately involved in the development of sustainable infrastructure worldwide. The International Development Association (IDA), in the period from July 1, 2014, to June 30, 2017 (IDA17), will place special emphasis on ensuring that development projects incorporate climate and disaster risk considerations and encompass a sharper focus on "value for money" through enhanced efforts to improve both results and cost-effectiveness (IDA 2014). In addition, in its 2013 report *Building Resilience: Integrating Climate and Disaster Risk into Development* (World Bank 2013a), the International Bank for Reconstruction and Development (IBRD) stressed that building climate resilience is critical for achieving the World Bank Group's goals of ending extreme poverty and building shared prosperity. The report called for the international development community to build long-term resilience, reduce risk, and avoid rising future costs.

This book supports these goals by providing practical solutions for those who will be involved in the planning, design, construction, and operation and maintenance of dams, reservoirs, and hydropower plants so that the threat of climate change and the need for sustainable, cost-effective infrastructure are taken into consideration. Specifically, this book addresses the critical threat of sedimentation—a process that reduces the storage capacity of reservoirs and with it all the water supply, flood control, and hydropower benefits they provide, and that damages hydromechanical equipment leading to a loss in hydropower generation.

The Importance of Sediment Management for Ensuring the Sustainability of Reservoir and Run-of-River Projects

Reservoirs are used worldwide to provide reliable water supply, hydropower, and flood management services. They are particularly important in areas of the world with high hydrologic variability, where the amount of water flowing in rivers varies significantly both seasonally and from year to year. In these areas, storing enough water for use during severe or multiyear droughts, and thereby ensuring the reliability of water and power supply, requires very large reservoir storage volumes. Countries where hydropower is an important source of energy often have both reservoir and run-of-river (ROR) projects. ROR projects, where preservation of storage is often a secondary objective, represent about 11 percent of all large dams (ICOLD 2015). For the remainder of dam projects, creating and maintaining reservoir storage is crucial to providing irrigation, water supply, flood control, multiple-use, and hydropower benefits.

Sedimentation poses a significant threat to the longevity, usefulness, and sustainable operations of both storage reservoirs and ROR projects (Palmieri et al. 2003). Over time, sediment builds up in reservoirs and displaces usable storage volume, which in turn negatively affects hydropower generation, reduces the reliability of domestic and irrigation water supply and flood management services, and degrades aquatic habitat. In ROR projects, sedimentation damages turbines and leads to inefficiencies in power generation and costly repair. In short, sedimentation is a major factor influencing the sustainability of dams, reservoir storage, and all types of ROR projects.

Dams have traditionally been designed under the "design life" paradigm, which entails estimation of the sedimentation rate and trap efficiency, and provision of a sediment storage pool volume equivalent to the design life (typically 50 or 100 years). Under this paradigm, consequences beyond the design life are not addressed, leading to decommissioning.[1] In many regions, however, new reservoirs are both costly and difficult to construct because the best (least costly) dam sites have already been used, and because there is intense resistance to the flooding of additional lands due to competing land uses and social and environmental concerns. The cost of dam decommissioning may also be very high. Finally, deposition of sediment in reservoirs removes it from downstream river reaches, thereby causing erosion of those reaches and degradation of aquatic habitat.

Therefore, as dams and reservoirs approach the end of their original design lives, most owners are interested in maintaining the infrastructure and continuing to generate economic and social benefits, including water supply, hydropower, and flood control, even if the benefits are not as large as in the original project. Extending the dam's life entails adopting a new design and operational paradigm that focuses on managing the reservoir and watershed system to bring sediment inflow and outflow into balance to the degree that doing so is practical, thereby giving the reservoir a greatly extended or even indefinite life.

Climate change is projected to increase hydrologic variability in many parts of the world, increasing the intensity of both floods and droughts. This variability will increase the need for larger reservoir volumes to ensure reliable water and power supplies and much-needed flood control. Climate change is also expected to increase sediment loads in many rivers, amplifying the threat of reservoir sedimentation. Therefore, it is essential that new dam and reservoir projects be designed, built, and maintained with the long-term threat of reservoir sedimentation in mind, and that existing projects be converted to sustainable use insofar as is possible. This perspective is consistent with the World Bank's efforts to develop climate screening tools.[2]

Solutions and Recommendations for Successful Sediment Management

The sustained threat of reservoir sedimentation and the anticipated increase in demand for large reservoir projects as a result of the effects of climate change oblige governments to assume a leadership role in sustainable development by investing in projects with lasting benefits and ensuring that investments made in the near term incorporate sediment management measures that will reduce future maintenance costs and ensure the long-term functionality of dams and hydropower infrastructure.

The World Bank often relies on client countries to hire consulting firms to address long-term sustainability of reservoir and ROR projects. However, it is the duty of World Bank staff to ensure that the consultants fulfill this role and that the goals of sustainable development are never compromised. This book calls on engineers and economists to incorporate sediment management measures into the early phases of project planning as part of a sustainable management approach. Specific sediment management techniques outlined herein include

- Reducing upstream sediment yield through erosion control and upstream sediment trapping,
- Managing flows during periods of high sediment yield to minimize trapping in reservoirs, and
- Removing sediment already deposited in reservoirs using a variety of techniques.

Purpose, Uses, and Organization of This Book

Purpose and Uses

The purpose of this book is twofold: (1) to illustrate why incorporating sediment management into dam projects is important and (2) to provide information on specific sediment management strategies that can be undertaken in projects as part of a sustainable sediment management approach. One of the key messages of this book is that incorporating sediment management into the planning and design phases of dam projects is essential for ensuring that the benefits of reservoir storage are sustained over the long term. Without sediment management, reservoir storage space is eventually lost, and it is extremely difficult—if not impossible—to reclaim it. Reservoir storage space is a key factor of production for water and renewable energy supply, and it is becoming increasingly important as climate change–related stresses increase and suitable storage sites become increasingly scarce. As a result, it is essential that projects incorporate sediment management at the outset as an integral part of their configuration to ensure lasting benefits.

This book aims to present techniques for sediment management in a manner that is accessible to a nontechnical audience. It is written primarily for World Bank Group team leaders, planners, government officials, and developers who are involved in the planning, design, construction, and operation and maintenance of dams, reservoirs, and hydropower plants. The book is neither an engineering manual nor an economic analysis manual; it is designed to fill the gap between general summaries, which are not useful in the practical sense when it comes to project planning and design, and detailed manuals, which are referenced throughout the book for further information.

This book is intended to inform readers of approaches to sustainable development of water resource infrastructure that will allow them to confidently review proposed projects. In particular, the Checklist for Sediment Management provided in appendix A highlights recommendations based on sediment problems that typically arise in projects. It is important to note, however, that sediment management strategies for specific projects must be tailored to site-specific conditions and limitations.

Organization

The remainder of this book is organized into nine chapters and an appendix:

- Chapter 2: Climate Change, Sediment Management and Sustainable Development discusses the importance of reservoir sediment management for preserving reservoir storage and illustrates how it contributes to satisfying the tenets of sustainable development.
- Chapter 3: Overview of Sedimentation Issues discusses the importance of reservoir storage and the impacts of reservoir sedimentation up- and downstream of dams.
- Chapter 4: Sediment Yield provides an overview of sediment yield, describes the important factors that determine the magnitude of sediment yield, and presents ways to estimate sediment yield.
- Chapter 5: Patterns of Sediment Transport and Deposition describes techniques for estimating the amount of sediment that will be deposited in a reservoir.
- Chapter 6: Sedimentation Monitoring discusses sedimentation monitoring procedures, bathymetric mapping of sediment, and estimation of sediment bulk density.
- Chapter 7: Sediment Management Techniques presents an overview of activities to combat reservoir sedimentation.
- Chapter 8: Sediment Management at Run-of-River Headworks describes basic concepts to consider in the design or rehabilitation of run-of-river headworks with regard to sediment management.
- Chapter 9: Reservoir Sustainability Best Practices Guidance summarizes sediment management strategies that will provide a higher level of assurance that the project operation can be sustained indefinitely.
- Appendix A: Checklist for Sediment Management is for use by project proponents to help ensure that projects adhere to the recommendations put forth throughout the book. The checklist is divided into three sections: sediment yield, sustainable sediment management measures, and sediment patterns and impacts.

Notes

1. Three examples are the San Clemente Dam (California, United States), the Matilija Dam (California, United States), and the Camaré (Pedregal) Dam (República Bolivariana de Venezuela).
2. World Bank Group. 2015. "Climate & Disaster Risk Screening Tools" (http://climatescreeningtools.worldbank.org).

References

ICOLD (International Commission on Large Dams). 2015. "Register of Dams." http://www.icold-cigb.org/GB/World_register/general_synthesis.asp.

IDA (International Development Association). 2014. "Report from the Executive Directors of the International Development Association to the Board of Governors 2014." Washington, DC.

Palmieri, A., F. Shah, G. W. Annandale, and A. Dinar. 2003. *Reservoir Conservation*: The RESCON Approach. Washington, DC: World Bank.

World Bank. 2013a. *Building Resilience: Integrating Climate and Disaster Risk into Development*. Washington, DC: World Bank. https://openknowledge.worldbank.org /handle/10986/16639.

———. 2013b. "Toward a Sustainable Energy Future for All: Directions for the World Bank Group's Energy Sector." World Bank, Washington, DC.

World Commission on Environment and Development. 1987. "Report of the World Commission on Environment and Development: Our Common Future." United Nations, New York.

CHAPTER 2

Climate Change, Sediment Management, and Sustainable Development

George W. Annandale

Introduction

Dam projects can generally be divided into run-of-river and storage projects. Run-of-river projects (figure 2.1, panel a), often used for hydropower generation, usually have small active storage volumes and large dead storage volumes. The objective is to maximize the head[1] and have just enough storage to satisfy peaking demands.

Storage projects (figure 2.1, panel b), in contrast, have large active storage volumes and small dead storage volumes. The active storage contains a large amount of water for irrigation and water supply and, in the case of flood management projects, is used to attenuate floods. Storage may also be used for hydropower generation. In such cases, the head used for power generation can vary, which will affect the efficiency of power production but increase the reliability of power supply (Annandale 2015).

Sediment management objectives in these two types of projects differ. For run-of-river projects, sediment management aims to improve operational efficiency. If sediment is not removed from run-of-river facilities before it enters the turbines, it may cause abrasion and clog the cooling water intakes of the electromechanical equipment, which increase operation and maintenance costs and diminish the amount of power that can be generated.

Sediment depositing in the dead storage space in run-of-river projects does not affect operational efficiency, although it may result in increased amounts of sediment entering the turbines. Sediment depositing in the active storage volume may diminish peaking ability, which, although undesirable, is often not addressed in project design (that is, designs have not historically allowed for removal of deposited sediment from the active storage).

The objective of sediment management in storage projects is to ensure project longevity for storing large amounts of water for use during droughts. Such storage

Figure 2.1 Active Storage Features of Run-of-River and Storage Reservoirs

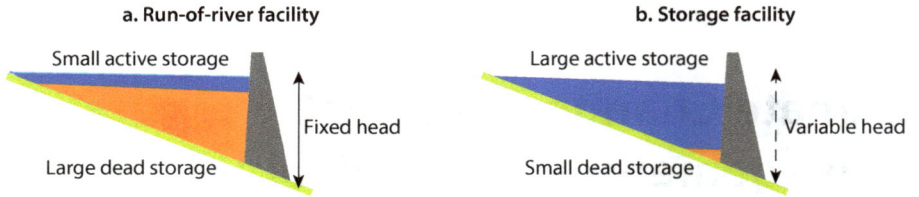

a. Run-of-river facility

Small active storage

Large dead storage

Fixed head

b. Storage facility

Large active storage

Small dead storage

Variable head

also provides the opportunity to attenuate floods. Failure in the past to design dams that include sediment management has resulted in the current net reduction of reservoir storage space worldwide. Annandale (2013) estimates that net reservoir storage has been decreasing since about 2000, and per capita storage worldwide is now at levels last seen in 1965.

In general, this book gives less attention to preserving storage space in run-of-river projects compared with storage projects. The reason is that reservoir storage is the most important function of dam projects, considering that run-of-river projects represent only 11 percent of all large dams (ICOLD 2015). Storage reservoirs deliver reliable water supply, irrigation, flood control, and hydropower services. Therefore, losing storage to sedimentation reduces the services provided by dams (Annandale 2013, 2015).

If storage is so important, why is reservoir sediment management to retain reservoir storage not routinely considered when designing dams? The answer is that most designers and economists rely on outdated design paradigms and are not familiar with modern reservoir sediment management techniques.

This book focuses on informing the reader of the basic concepts underpinning sustainable sediment management strategies. The purpose is to emphasize the importance of managing sediment to prevent or minimize storage loss. The book lists and categorizes reservoir sediment management techniques to encourage changes to design paradigms. It does not provide detailed instructions on how to design reservoir sediment management systems, nor on how to execute economic analysis. The key principles of integrated engineering and economic analysis of sediment management are well covered elsewhere; in particular, see Palmieri et al. (2003), which is extensively referenced here. Economic analysis is critically important, and a discussion of basic concepts is included for the nonspecialist reader.

Discussion of the economic analysis of sediment management leads to one of the fundamental points of this book: without sediment management, dam projects run the risk of exhausting prime water storage sites, even as the risk of higher sediment loads is likely to accelerate as climate changes over the next 30–100 years. Yet with sediment management, these storage sites, a key factor of production for cost-effective water and energy supply, can be maintained as a sustained resource for long periods, often in excess of 100 years (in some cases, in perpetuity).

The complexities of natural resource economics have been the subject of intense inquiry since the publication of Hotelling's paper in 1931. In spite of numerous contributions made in this field of investigation, common agreement on some economic analysis procedures is still lacking, particularly as it relates to intergenerational equity. The concept of intergenerational equity is a key element of the reservoir conservation (RESCON) approach put forward in Palmieri et al. (2003), and derives in part from the concept of sustainable development.[2] The United Nations appointed the World Commission on Environment and Development in 1987 to investigate sustainable development, and the commission made recommendations on how to achieve it—its widely quoted definition (see chapter 1) reflects an appropriate focus on the needs and aspirations of humans, and its recommendations address the importance of striking a balance between current and future needs. More recently, Denton et al. (2014), as part of the Intergovernmental Panel on Climate Change's Fifth Assessment process, have updated the thinking on sustainable development and identified climate change as a key threat, expanding the concept of sustainable development to reflect the need for climate-resilient development:

> Improved understandings of the short- and long-term implications of climate change and extreme events...have influenced conceptualizations of sustainable development and related objectives such as poverty reduction, health, livelihood and food security, and other aspects of human welfare related to the idea of "climate-resilient development." (Denton et al. 2014, 1108).

Hydropower and water supply, and water storage infrastructure more generally, are appropriate focal points for both sustainable development and climate resilience, playing an important role in both adaptation and mitigation agendas for climate change. In turn, sediment management is a necessary element of any sustainable and climate-resilient plan that includes hydropower and reservoir storage.

The Dual Nature of Reservoir Storage

When considering sustainable development of reservoir storage, which is a key factor of production for water and renewable energy supply, storage space should be considered a natural resource that is created by the presence of a dam. Storage space can be classified as either a renewable or an exhaustible resource, depending on how the dam and reservoir are designed and operated. If such a system is designed and operated in a way that allows the reservoir to fill with sediment, the designers have treated it as an exhaustible resource. However, if designed and operated to either prevent or minimize storage loss due to reservoir sedimentation, the system is being treated as a renewable resource.

It is important to note that the classification of reservoir storage space as either renewable or exhaustible is a choice and is a deliberate decision made by owners, engineers, economists, financiers, and operators. If these actors decide to maintain reservoir storage space through implementation of reservoir sedimentation, management, such a reservoir could potentially be used in perpetuity, thereby satisfying the tenets of sustainable development. If, however, the decision is made to allow a reservoir to fill with sediment without any intervention, a deliberate decision has been made by the development team to treat the reservoir as an exhaustible resource.[3]

Shifting Paradigms

Developing and retaining enough reservoir storage space to satisfy global needs over the very long term requires inclusion of reservoir sediment management facilities in dam and reservoir designs right from the start, at project conception. It requires abandoning the conventional design life approach to dam design and adopting a life-cycle management approach. This change has implications for the economic analysis of projects. Although many World Bank project plans now incorporate a life-cycle perspective (see, for example, box 2.1 on the Dasu Dam project in Pakistan), historically that perspective has been absent. Implementing a life-cycle management approach demands consideration of how many years of benefits and costs should be included in the economic analysis. It also affects the selection of discount rates to account for intergenerational features of development.

Box 2.1 Sediment Management in the Dasu Hydropower Project, Pakistan

The Dasu Hydropower Project (DHP) is a 4,320 megawatt run-of-river facility to be constructed on the Indus River, about 240 kilometers upstream of Tarbela Dam. The Indus River is known for its high sediment loads, estimated to be on the order of 200 million tons per year at the project location. Another project, Diamer Basha Hydropower Project (Basha), may be constructed upstream of DHP at a later stage. Should this happen, the sediment inflow into DHP is expected to decline to about 45.6 million tons per year.

Preservation of reservoir storage volume and protection of hydraulic machinery from abrasion by sediment required design of reservoir sediment management facilities. Based on the assumption that Basha might not be constructed, the DHP project is equipped with nine 6.4-meter diameter low-level outlets in the dam and two 9.4-meter-equivalent diameter flushing tunnels in the right abutment, which can jointly be used to implement drawdown flushing. Combined, these outlets can freely discharge 4,400 cubic meters per second of water to remove deposited sediment.

Should Basha not be built, the designers recommend drawdown flushing for one month every year, commencing in year one. The design calculations demonstrate that drawdown flushing not only accomplishes storage preservation goals but also reduces wear and tear on the turbines from abrasion by sediment. Successful sediment management using this approach leads to the estimate that the repair cycle for turbines (to deal with the effects of abrasion) will be very long; on the order of 16–24 years.

box continues next page

Box 2.1 Sediment Management in the Dasu Hydropower Project, Pakistan *(continued)*

A less beneficial approach would be to commence drawdown flushing 15 years after project commissioning, and then repeat the flushing every year, or once every 3–5 years depending on practical outcomes at that time. The latter approach, that is, commencing drawdown flushing after 15 years flushing either and every year or every 3–5 years, was found to increase the frequency of repair cycles for turbines to about every four years by the time flushing commences after 15 years. It is deemed beneficial to commence drawdown flushing immediately after commissioning of the dam and repeat it annually. Operational costs associated with turbine repair increase about four- to sixfold when delaying commencement of sediment management.

The economic rate of return (ERR) of the Dasu Phase I project is not much affected by reservoir sediment management. As shown in table B2.1.1, in the absence of sediment management, the ERR is 25 percent. If sediment management is started immediately (in year one) and if drawdown flushing occurs annually, the ERR drops to 20.8 percent. If flushing commences after 10 years, the ERR drops to 24.4 percent, while it remains at 25 percent if flushing commences after 15 years.

Table B2.1.1 Economic Rate of Return (ERR) for Dasu Project Phase I

Commencement of flushing	Frequency of flushing	ERR of Dasu Project (%)
No flushing	not applicable	25.0
Year 1	Annually	20.8
After 10 years	Annually	24.4
After 15 years	Annually	25.0

However, it is important to note that, although this analysis incorporates the cost of sediment management within the intragenerational period, it effectively does not reflect a significant benefit for implementation of sediment management over time, if only because it uses a relatively high discount rate (10 percent) consistent with former World Bank guidance rather than the more recent recommendation of 6 percent. Furthermore, the data presented here for Dasu do not reflect a sensitivity analysis using a declining discount rate for benefits that accrue after 30 years, a test that would be reasonable for a long-lived asset such as a hydropower facility.

The Water and Power Development Authority of Pakistan selected the alternative wherein drawdown flushing commences after 15 years, which provides "adaptive management" flexibility.

Designing and constructing a dam with the required facilities in place allows for sediment management should Basha not be built. And in the long term, even if Basha is built, sediment management will be possible as DHP gradually fills with sediment and flushing operations at Basha demand sediment pass-through.

This section briefly presents and contrasts the characteristics of customary engineering design philosophy (that is, the design life approach) and the recommended life-cycle management approach. The life-cycle management approach provides the setting for sustainable development.

Design Life Approach

Conventional civil engineering design philosophy embraces the concept of a "design life," after which the infrastructure is simply exhausted. In essence, this concept means that infrastructure will serve its purpose for a finite period. If the present value of benefits obtained from the infrastructure during that period is greater than the present value of the costs, the infrastructure is deemed economically viable. For convenience the term "design life" will be used to identify this finite period.

The design life approach is a linear thinking process, starting with planning and design and proceeding to construction, operation, and maintenance of the infrastructure, and finally its disposal (figure 2.2). More often than not, the design life approach works well for conventional civil infrastructure, such as roads, bridges, and buildings—but not for dams and reservoirs.

This approach works well for conventional civil engineering infrastructure because this type of infrastructure is easily refurbished at the end of its design life and is usually not disposed of. For example, the design life of a road may be, say, 30 years. At the end of 30 years the road can be resurfaced and upgraded fairly easily for continued use. As a result, in spite of the assumption of a finite design life, conventional civil infrastructure can be used sustainably through regular refurbishment at a fraction of the original cost of construction.

The ease of refurbishment of conventional civil infrastructure justifies the use of the design life concept. However, when considering dams and reservoirs, this thought process is often not defensible. The problem with applying the design life approach to dams and their reservoirs is that, once storage reservoirs, particularly large ones, are filled with sediment, they often cannot be used anymore. At that point, the reservoir as a resource has been exhausted and lost unless extreme measures can be taken—and the result is that some of the world's best reservoir sites can only be replaced by sites with lesser location and engineering advantages.

Removal of deposited sediment is no simple task. The volume of deposited sediment in a reservoir over its design life can amount to millions, if not billions, of cubic meters. For example, the net cost of decommissioning the Tarbela Dam according to one estimate reported in Palmieri et al. (2003) is US\$2.5 billion.

Therefore, the design life approach, although feasible for most conventional civil infrastructure, is generally not suitable for designing dams and their reservoirs.

Figure 2.2 Design Life Approach to Infrastructure Design

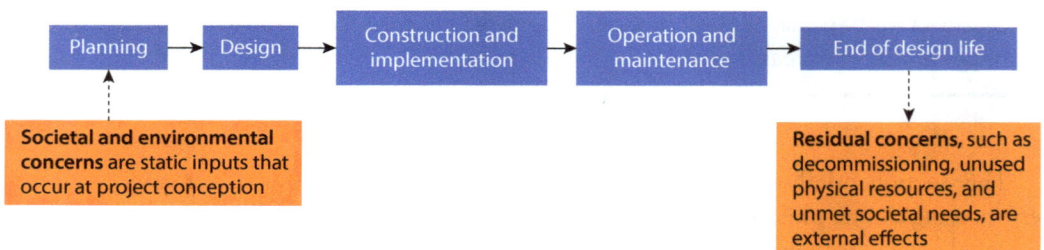

Source: Adapted from Palmieri et al. 2003.

The pursuit of sustainable development for dams and their reservoirs requires a new way of thinking. The challenge is to adopt a project development approach that will result in continuous and perpetual use of dams and reservoirs.

Life-Cycle Management Approach

The desired paradigm shift that would facilitate sustainable development of dams and reservoirs can be accomplished by adopting a life-cycle management approach, a concept that has been incorporated in the RESCON approach developed for the World Bank (Palmieri et al. 2003) (also see box 2.2). It commences with planning, design, and construction phases, as before. However, once the

---

Box 2.2 A Note on Terminology

Dams used for hydroelectric power generation frequently (though not always) use reservoir storage to improve the reliability of power production. Although hydropower is typically considered a renewable resource, it becomes nonrenewable when sedimentation displaces reservoir storage, resulting in what is often an irreversible loss. In addition, reservoir storage can be seen as a nonrenewable resource because the number of dam sites is limited. As a result, preserving reservoir storage is critically important for sustainable development worldwide for the benefit of our own and future generations.

The life-cycle management, sustainable use approach as applied in sections of this book derives principally from the reservoir conservation (RESCON) approach articulated in Palmieri et al. (2003). The introductory material in the RESCON manual uses the terms "life-cycle management" and "sustainable use" to define this approach. These terms, while not exactly interchangeable, are meant to convey a similar perspective:

> Common engineering practice uses a "design life" approach in dam and reservoir design, which assumes that over the course of its life, a water resource project would recover investment costs through the benefits generated by the project. This approach does not take into account what happens to the project at the end of its design life, and it is assumed that problems with reservoir sedimentation and eventual retirement will be addressed by future generations. The "life cycle management" approach advocated by this book instead aims at designing and managing water resource infrastructure for sustainable use. This approach requires the incorporation and use of sediment management facilities. (Palmieri et al. 2003, vii)

The concept of life-cycle management as used in this book conveys the principles of sustainable use and sustainable development. Sustainable development in the World Bank's conception effectively adopts one of the most common definitions of this term: "Development that meets the needs of the present without compromising the ability of future generations to meet their own needs" (World Commission on Environment and Development 1987, 43). It is this approach that the authors of this volume hope will be adopted by all owners, operators, and managers of hydropower and water storage facilities with respect to sediment management.

---

infrastructure has been built, the concept adopts a circular nature and abandons the concept of disposal.

The life cycle of dams and reservoirs consists of operation and maintenance, continued and regular implementation of reservoir sediment management approaches, and regular refurbishment of the dam and appurtenant structures (figure 2.3). Reservoir sediment management and refurbishment allow for continued use of the dam and its reservoir, ideally in perpetuity. In principle, the approach does not include the element of disposal.

A major difference between the life-cycle management approach and the design life approach is the focus on preventing storage loss caused by reservoir sedimentation. It eliminates the threat of losing the reservoir's ability to store water over the very long term and promotes continued use of the dam and reservoir, providing utility to both current and future generations.

It stands to reason that adoption of the life-cycle management approach requires a different attitude toward engineering planning and design and toward economic analysis. From an engineering point of view, consideration of how sediment might be managed over the very long term becomes important, particularly as it relates to preserving reservoir storage. For new dams, it means conducting detailed investigations into how storage loss due to reservoir sedimentation might be avoided in the future and incorporating the required facilities into the

Figure 2.3 The Life-Cycle Management Approach

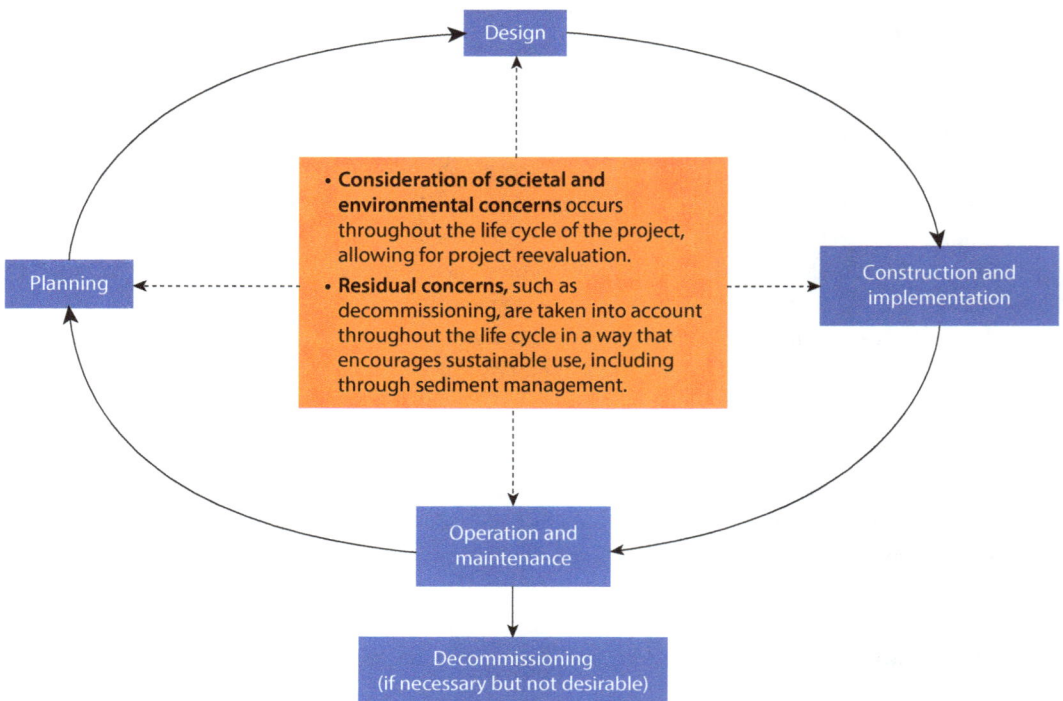

Source: Adapted from Palmieri et al. 2003.

design right from the start to ensure the goal is accomplished. From an economic analysis point of view, the selection of the period of analysis and of discount rates becomes important.

Economic Analysis and Sustainable Development

This section focuses on distinguishing between basic principles of conventional economic analysis and principles that are relevant when using the life-cycle approach.

Conventional Approach

A project economic analysis begins with an assessment of the economic flows, which is then further analyzed to quantify the net present value (NPV) and internal rate of return. These measures are typically calculated in a World Bank Project Assessment Document, and are also referred to in the RESCON tool (Palmieri et al. 2003).

A simplified cash flow of a dam and reservoir project is illustrated in figure 2.4. The results are for the PB Soedirman reservoir, constructed in 1988 and 1989 in Central Java, Indonesia. The initial project construction cost is shown on the left of the figure as a negative cash flow value of $126 million. Annual costs for operation and maintenance of the project are $1.26 million per year and shown for each of the 35 years of the expected lifetime of the facility. The final decommissioning costs, shown at the end of the project lifetime, are $50 million.

The benefits of constructing the facility accrue during the full 35-year project lifetime, and are shown as positive cash flow. This initial illustration reflects a design life approach, and also reflects the reduced hydropower generation

Figure 2.4 Standard Approach to Economic Analysis of Dams and Reservoirs, PD Soedirman Reservoir, Indonesia

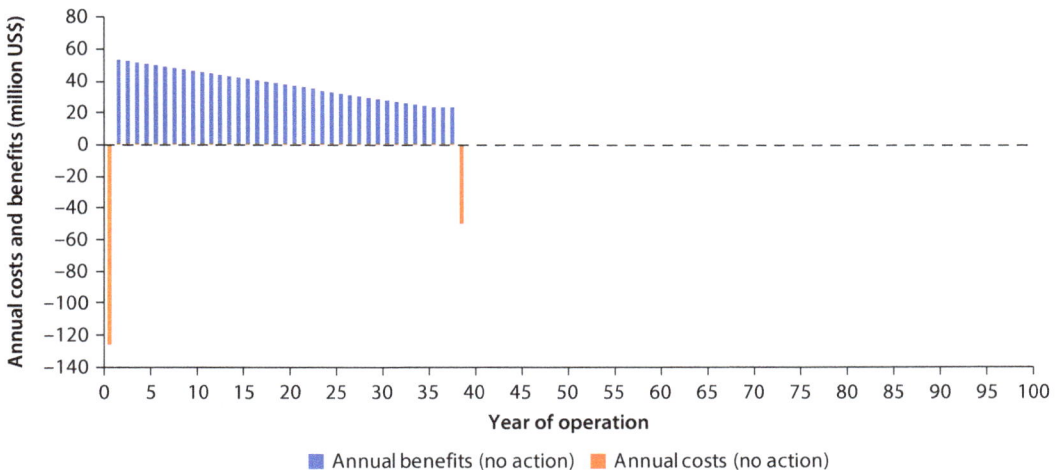

Box 2.3 Tarbela Dam, Pakistan

Tarbela Dam is a major facility that was originally designed without considering the use of reservoir sediment management to preserve its storage over the long term. The fifth periodic inspection of the dam found that it could have originally been designed to regularly remove sediment from the reservoir using drawdown flushing (Annandale 2008). Subsequent development of the river downstream of the dam—barrages, irrigation turnouts, and other infrastructure—now precludes such releases of sediment. The amount of sediment already deposited in the reservoir is so large that its removal and storage poses an almost insurmountable problem.

Construction of Tarbela Dam, located in the Indus River, Pakistan, was completed in 1974. The facility, which is one of the largest dams in the world, is primarily used to supply water for irrigation, and secondarily to generate hydroelectric power. Tarbela Dam supplies 30 percent of the country's irrigation water and 30 percent of its electric energy.

Sedimentation in Tarbela Dam's reservoir is severe and has been a concern since commissioning. The average annual sediment discharge into the reservoir is about 181 million tons. The trap efficiency of the reservoir, that is, the percentage of incoming sediment retained by the reservoir, is greater than 95 percent in most years. The original gross reservoir capacity at commissioning was 14.33 billion cubic meters, which declined to 10.105 billion cubic meters by 2006. This is a reduction of 29.48 percent in 36 years, that is, a reduction rate of about 0.8 percent per year. The live storage in the reservoir has decreased from 11.939 billion cubic meters in 1974 to 8.550 billion cubic meters in 2006, a reduction of 28.28 percent (Annandale 2008). Continued reduction in live storage space is a concern because it will eventually result in inadequate availability of irrigation water and a subsequent reduction in power supply.

capacity over time caused by the effects of reservoir sedimentation (note the gradual decline in project benefits over the 35-year lifetime).

The resulting loss of generation and other reservoir project services is further illustrated by the story of the Tarbela Dam, in box 2.3. Tarbela, like PB Soedirman, was built without regard to sediment management.

Benefits and Costs for Future Generations

The illustration in figure 2.4 does not yet incorporate calculation of the NPV, which requires selection of an appropriate discount rate that reflects the value of intergenerational equity. This topic is discussed extensively in the literature (see, for example, Hotelling 1931; Clark 1973; Goulder and Williams 2012; Arrow et al. 2013). Some of the common threads in these papers are the recognition of two types of objectives, two types of discount rates, and the potential benefit of and justification for using a declining discount rate. For example, Goulder and Williams (2012) argue that the selection of a discount rate depends on the objective of the economic analysis. The authors identify two types of objectives: those intended to augment social welfare and those aimed at achieving a net financial benefit for all. The discount rate associated with maximizing financial return is generally known as the investment-based (or finance-based) discount rate, while that associated with augmenting social welfare is known as the consumption-based discount rate. The consumption-based discount rate is the rate at which society is willing to trade consumption in the future for consumption today.

The World Bank has traditionally used rates in the 10–12 percent range for project analyses, consistent with financial market rates of interest. Recent guidance from a World Bank group convened to recommend discount rates for use in World Bank projects recommends that a discount rate of 6 percent be the new default rate and that the full benefits and costs for projects be calculated; discounting calculations should not be terminated at 20 or 30 years, or some other arbitrary cutoff date (Fay et al. 2016). Most reservoir projects would be classified as long-lived, that is, having the potential to provide benefits for 100 years or more.

At this time, economists do not universally agree on desirable discount rates (see, for example, Campos, Serebrisky, and Suárez-Alemán 2015); declining discount rates are used as one of the options to place a value on creating intergenerational equity (for example, France and the United Kingdom use declining rates [see, for example, OECD 2007]).

For the PB Soedirman example, a declining discount rate, as shown in table 2.1, was used to illustrate the benefit of using such a rate. Table 2.2 illustrates the effect on the NPV of the project when using alternative discount rates. As indicated in the table, a higher discount rate lowers the NPV, and use of a declining discount rate for this asset with a useful life of 35 years raises the NPV relative to either constant discount rate. The table highlights as well that incorporating a decommissioning cost also lowers the NPV but, because of the effect of discounting over time, by an amount less than the expected $50 million decommissioning cost. In the case with no sediment management, the inclusion of a decommissioning cost might be reasonable because sediment accumulation

Table 2.1 Recommended Declining Discount Rate Sequence

Period (years)	0–30	31–75	76–125	126–200	201–300	301+
Discount rate (% per year)	6.0	4.5	3.5	3.0	1.5	1.0

Table 2.2 The Effect of Alternative Discount Rates on PB Soedirman Project Net Present Value Using the Conventional Design Life Approach

	Net present value (US$ million)	
Discount rate	Approach without sediment management and without decommissioning cost	Approach without sediment management but including decommissioning cost
Constant discount rate of 12%	268	239
Constant discount rate of 6%	478	473
Declining discount rate	492	482

Extending the Life of Reservoirs • http://dx.doi.org/10.1596/978-1-4648-0838-8

would result in the reservoir filling at the end of its conventional useful life, prompting a decommissioning process.

Acknowledging the Cost of Lost Storage Space

The remaining chapters of this book show that reservoir storage space is required to provide reliable flood management, power, and water supply services. The demand for reservoir storage space is likely to increase as the effects of climate change set in. Increased hydrologic variability due to climate change will result in greater demand for reservoir storage, and the fact that it is in limited supply increases the importance of preserving storage space (Annandale 2013). This value of storage, while conceptually appealing, does not readily enter the cash flow calculus for the PB Soedirman example. In theory, scarcity of high-quality reservoir sites would be reflected in the opportunity cost of services provided by the reservoir—in other words, if other water and electric energy supply sites are more expensive than the reservoir currently in use, then the value of the services provided by that site ought to be higher. In practice, estimating the value of storage is quite difficult. Consequently, in economic analyses of this type, no allowance for a direct measure of the value of reservoir storage itself is made, only of the services the reservoir offers.

An Alternative Life-Cycle Approach

Another approach is outlined in this section for two additional scenarios that incorporate sediment management, consistent with the recommended life-cycle approach to reservoir management.

Sediment management

Implementation of reservoir sediment management approaches that prevent storage loss introduces an additional cost. In the case of the PB Soedirman Reservoir, a sediment management system consisting of a sediment bypass requires an initial investment of $20 million, operation and maintenance cost of $400,000 each year, and a loss of income during 5.4 months each year when the bypass is in effect (it is assumed that no power is generated during this period).

In the example presented below, three differences can be noted when compared with figure 2.4. First, the initial costs of a sediment bypass retrofit are incurred in year 35. Second, the sediment management retrofit extends the life of the facility to 100 years, providing benefits in years 35–99 and postponing the decommissioning cost to year 100. Third, the measure implies a hydropower production penalty in each operating year from 35 to 99, as can be seen by the lower annual benefits for those years compared with year 34.

The NPV implications of this investment are shown in table 2.3, which has two more columns than table 2.2. The third column in the table provides the NPV results for the sediment bypass option cash flow results shown in figure 2.5. As indicated in the table, at a 12 percent discount rate, the net benefits of sediment management are negligible—they do not change the NPV relative to when no sediment management is implemented.

Table 2.3 The Effect of Alternative Discount Rates and Sediment Management on Net Present Value of PB Soedirman Project

	Net present value (US$ million)			
Discount rate	Approach without sediment management	Approach without sediment management including decommissioning	Approach with sediment management including decommissioning after 100 years of operation	Approach with sediment management including decommissioning after 200 years of operation
Constant discount rate of 12%	268	239	239	239
Constant discount rate of 6%	478	473	482	482
Declining discount rate	492	482	509	520

Figure 2.5 Life-Cycle Approach Reflecting Sediment Management Investments

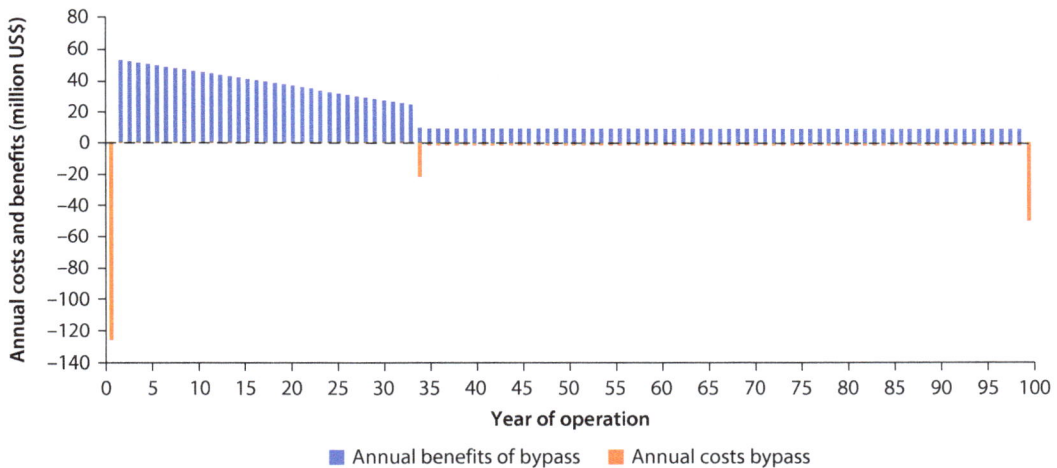

However, at the lower 6 percent discount rate, or when using a declining discount rate, the NPV increases relative to the NPV with no sediment management.

A second scenario is shown in the last column—in this scenario, the facility continues to operate through a 200th year with the sediment bypass investment. Comparing the third and fourth columns, it can be seen that almost no additional economic benefit derives when using a constant discount rate (either 12 percent or 6 percent). However, when using a declining discount rate, applicable for a long-lived, 200-year asset, the NPV increases to $520 million from $509 million. The declining discount rate reflects a greater value placed on intergenerational equity, resulting in an increase of $281 million compared with the calculation using a constant 12 percent discount rate.

Evaluation of sediment management with regard to its effect on cash flow is already being incorporated into some World Bank studies, as described in box 2.1 on the Dasu Dam.

Summary

Current design and economic analysis paradigms consider benefits and costs over a finite period, known as the design life. Although this approach is defensible for some civil infrastructure such as roads and bridges, it is not appropriate for designing dams and determining their economic value. When designing and operating dams, a life-cycle management approach is appropriate because of the unique characteristics of reservoir storage space.

Reservoir storage space created by dams is a natural resource with a dual character; it can either be exhaustible or renewable depending on the developer's decisions. If a reservoir is allowed to fill with sediment unhindered, it is deliberately and consciously classified as an exhaustible resource. However, if designed with reservoir sediment management and storage preservation in mind, the storage space is consciously classified as a renewable resource.

In contrast with the design life approach, the life-cycle approach allows sediment management technology in new dam designs to be considered right from the start. Applying the same concept to existing dams and reservoirs requires refurbishment to allow for reservoir sediment management to promote sustainable development. Designing new dams in accordance with the life-cycle management approach and refurbishing existing dams with this approach in mind aims at, in theory at least, using these facilities in perpetuity and thereby respecting the tenets of sustainable development.

Notes

1. The head of a hydropower project is the elevation difference between the water surface elevation in the upstream reservoir and the water surface elevation of the downstream river. The pressure head is used to generate hydroelectric power.

2. The original 2003 version of RESCON is currently being updated by the World Bank, and is scheduled to be released in 2016.

3. It should be noted that sustainable development is most often defined not at the project scale, but at a societal scale. The two key reasons are that (1) project-level optimization necessarily leaves out the optimization of available substitutes for that resource and (2) project-level optimization will usually result in a finding that some version of the project should move forward, but rarely results in full abandonment of the project, even if society might be better served by another project or source of economic welfare. The key assumptions in this book are that water will be most cost-effectively supplied through storage sites, and that good storage sites are rare and exhaustible. Under these conditions, a sediment-filled reservoir can be considered an exhausted resource for water supply.

References

Annandale, G. W. 2008. "Engineering and Hydrosystems Report: Tarbela Dam Fifth Periodic Inspection: Reservoir Sedimentation Management." Submitted to WAPDA, Islamabad, Pakistan.

———. 2013. *Quenching the Thirst: Sustainable Water Supply and Climate Change.* Charleston, SC: CreateSpace.

———. 2015. "Policy Considerations for Sustainable Hydropower—Reliability, Climate Change and Sedimentation", HYDRO 2105, Bordeaux, France.

Arrow, K., M. Cropper, C. Gollier, B. Groom, G. Heal, R. Newell, W. Nordhaus, R. Pindyck, W. Pizer, P. Portney, T. Sterner, R. S. J. Tol, and M. Weitzman. 2013. "Determining Benefits and Costs for Future Generations." *Science* 341: 349–50.

Campos, T., T. Serebrisky, and A. Suárez-Alemán. 2015. *Time Goes By: Recent Developments on the Theory and Practice of the Discount Rate.* Washington, DC: Inter-American Development Bank.

Clark, C.W. 1973. "The Economics of Overexploitation." *Science* 181: 630–34.

Denton, F., T. J. Wilbanks, A. C. Abeysinghe, I. Burton, Q. Gao, M. C. Lemos, T. Masui, K. L. O'Brien, and K. Warner. 2014. "Climate-Resilient Pathways: Adaptation, Mitigation, and Sustainable Development." In *Climate Change 2014: Impacts, Adaptation, and Vulnerability. Part A: Global and Sectoral Aspects. Contribution of Working Group II to the Fifth Assessment Report of the Intergovernmental Panel on Climate Change*, edited by C. B. Field, V. R. Barros, D. J. Dokken, K. J. Mach, M. D. Mastrandrea, T. Bilir, M. Chatterjee, et al., 1101–31. Cambridge, U.K.: Cambridge University Press.

Goulder, L.H., and R. C. Williams. 2012. *The Choice of Discount Rate for Climate Change Policy Evaluation.* Washington, DC: Resources for the Future.

Fay, M., S. Hallegate, A. Kraay, and A. Vogt-Schilb. 2016. "Discounting Costs and Benefits in Economic Analysis of World Bank Projects." Unpublished, World Bank, Washington DC.

Hotelling, H.M. 1931. "The Economics of Exhaustible Resources." *Journal of Political Economy* 39: 137–75.

ICOLD (International Commission on Large Dams). 2015. "Register of Dams." http://www.icold-cigb.org/GB/World_register/general_synthesis.asp.

OECD (Organisation for Economic Co-operation and Development). 2007. "Use of Discount Rates in the Estimation of the Costs of Inaction with Respect to Selected Environmental Concerns." OECD, Paris.

Palmieri, A., F. Shah, G. W. Annandale, and A. Dinar. 2003. *Reservoir Conservation: The RESCON Approach.* Washington, DC: World Bank.

CHAPTER 3

Overview of Sedimentation Issues

George W. Annandale

Introduction

Reservoir sedimentation occurs when sediment carried by a river flowing into a reservoir is deposited in the reservoir upstream of a dam. The sediment carried by the inflowing river is deposited in a reservoir because the water slows down after entering it and no longer has the ability to transport the sediment. Such deposits consume reservoir storage space that was originally intended for water storage, thereby impeding the intended function of the dam and reservoir. Sediment deposition in reservoirs also leads to smaller amounts of sediment being released to river reaches downstream of dams, which results in changes to river morphology, degradation of the downstream river channel and aquatic habitat, and reduction of food sources consumed by fish in rivers downstream of dams. This chapter discusses the importance of reservoir storage and the impacts of reservoir sedimentation up- and downstream of dams. The importance of reservoir sediment management is emphasized, and the global impact of reservoir sedimentation is presented.

The Importance of Storage

The non-uniform flow in rivers, both seasonally and interannually, results in shortages of water for hydropower generation and water supply during low-flow periods. Dams constructed across rivers provide storage space to capture water during times when flow is high for use during times when flow is low, thereby increasing the reliability of water and power supply.

The required reservoir storage volume depends on hydrologic variability. If the amount of water flowing in a river does not change significantly from year to year, the required reservoir storage space is relatively small, only providing enough volume to bridge shortages during within-year, seasonal low flows. The reservoir volume required to reliably supply water and power during multiyear droughts is much larger.

The duration of multiyear droughts is positively correlated with the annual coefficient of variation of river flow.[1] Globally, the annual coefficient of variation of river flow generally ranges between about 0.2 and 0.8 and higher (Annandale 2013). When the annual coefficient of variation is on the low side (0.2), the river is characterized by annual flow volumes that are roughly the same from year to year. The principal variation in those stream flows occurs only seasonally, within the year. In contrast, if the annual coefficient of variation is high (0.6 to 0.8 or higher), the river flow is characterized by multiyear droughts occurring in a cyclical manner. Such droughts can last from two to seven or more consecutive years, during which time annual flows are lower than the long-term mean annual flow (MAF). Map 3.1 provides an illustration of world regions where multiyear droughts may occur on a regular basis. These regions require dams with very large storage spaces (Annandale 2013).

To reliably supply water during such long, multiyear droughts, very large reservoir storage spaces are necessary. Large floods that occasionally occur in the regions identified in map 3.1 are captured by these large reservoirs, storing enough water for use during the multiyear droughts, thereby increasing the reliability of water and power supply.

Figure 3.1 provides the relationship between the required reservoir volumes that will supply water at 99 percent reliability for varying hydrologic variability, based on the Gould-Dincer Method and assuming that annual flows can be described by the log-normal probability distribution (McMahon et al. 2007). Figure 3.1 includes a dotted line separating run-of-river

Map 3.1 World Regions Where Multiple-Year Droughts Occur

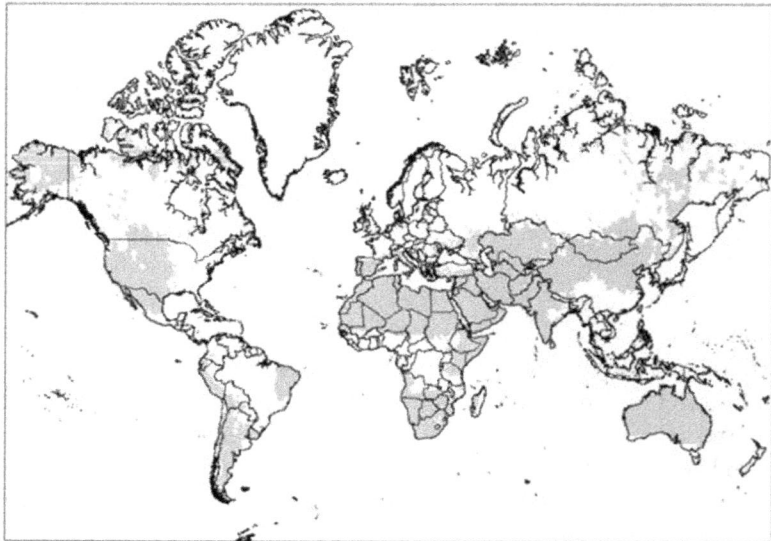

Source: Annandale 2013. © George W. Annandale. Used with permission. Reuse may require further permission.

Figure 3.1 Relationship between Yield and Hydrologic Variability at 99 Percent Reliability

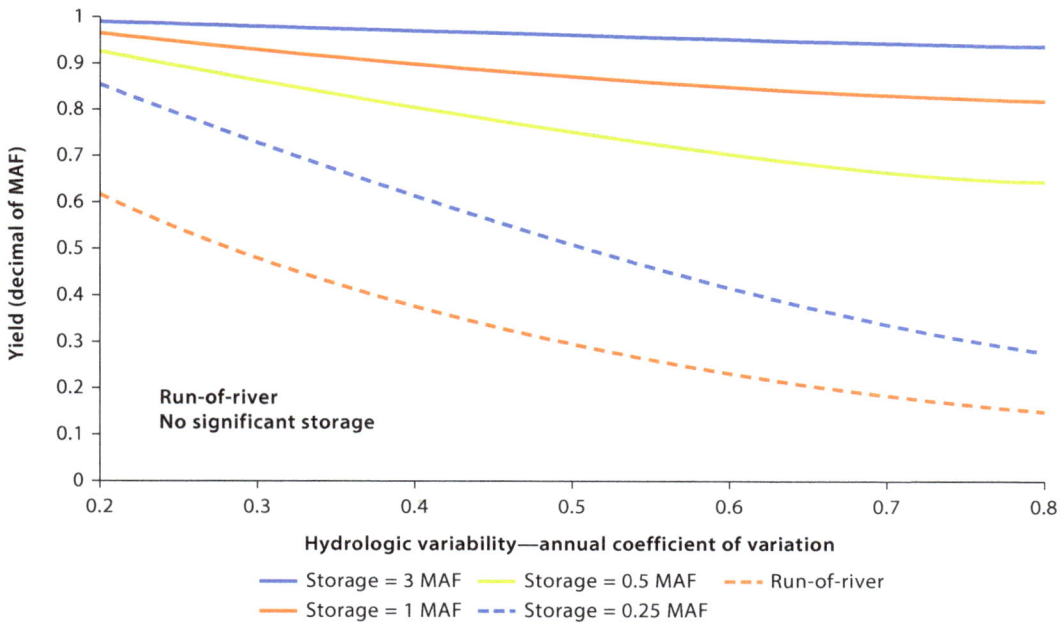

Source: Based on McMahon et al. 2007
Note: MAF = mean annual flow.

(below the dotted line) and storage reservoirs.[2] Run-of-river reservoirs are located in regions where annual flows do not differ much from year to year, that is, where only very small reservoir volumes may be required to bridge within-year, seasonal low flows.

The region in the figure above the dotted line contains a number of curves representing various reservoir volumes. The reservoir volumes are expressed as a ratio of the MAF. For example, a reservoir volume of 0.5 MAF equals half the MAF volume in the river where it is located.

The yield from a reservoir is presented on the vertical axis of the figure, also expressed as a percentage of the mean annual river flow volume.[3] Note that large reservoir volumes provide greater amounts of water at 99 percent reliability than smaller reservoirs. This is particularly true if the hydrologic variability (coefficient of variation) is high.

Figure 3.1 aptly illustrates the importance of preserving reservoir storage space by countering the effects of reservoir sedimentation, notably in regions with high hydrologic variability (that is, semi-arid and arid regions). Reservoir sedimentation reduces reservoir storage space, thereby reducing the amount of water that can be supplied at a specified reliability. For example, if the hydrologic variability is 0.6, and reservoir sedimentation reduces the storage volume in a reservoir from 0.5 MAF to 0.25 MAF, then the yield would decrease from about 70 percent to 40 percent of the MAF.[4]

Reservoir storage is also required for flood control. When large floods flow into reservoirs, the water is temporarily stored. To attenuate the flood, the water that has been temporarily stored in the reservoir is released downstream at a slower rate. The reservoir storage volume required for flood control is determined through detailed studies.

It is generally agreed that hydrologic variability will increase in the future as a result of climate change (IPCC 2013; Rahman et al. 2015). Increased annual hydrologic variability points to the occurrence of longer, multiyear droughts and, therefore, a greater need for large reservoir storage volumes to reliably supply water and power (Annandale 2013, 2015). Increased hydrologic variability will also result in larger floods, which will require larger storage volumes to attenuate those floods. Storage volume loss due to reservoir sedimentation is, therefore, undesirable. Reservoir sediment management techniques, which will become increasingly important as climate change sets in, are discussed starting in chapter 7.

Sedimentation Impacts Upstream of a Dam

Storage Loss and Its Impacts

Knowledge of the way in which sediment deposits in a reservoir, which is the subject of chapter 4, is essential in determining the impacts of sedimentation on the performance of dam and reservoir projects. Figure 3.2 provides a general impression of typical upstream impacts, indicating that total reservoir storage space is the sum of active and dead storage space. Dead storage space is located below the low water level determined by the elevation of the lowest outlet, while active storage space contains the water that may be released for power generation or for water supply, or may, in the case of flood control dams, be reserved for flood management.

Figure 3.2 Storage Loss in Active and Dead Storage Zones Due to Reservoir Sedimentation

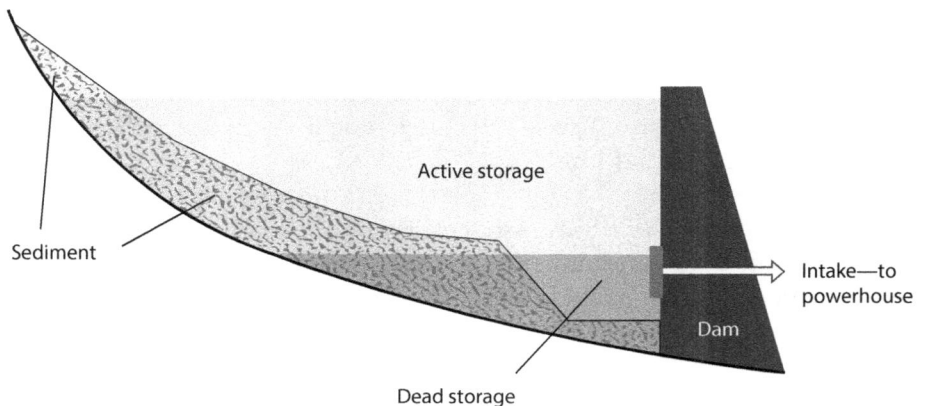

Sediment deposition frequently commences on the upstream side of a reservoir and gradually moves downstream in the shape of a delta.[5] Some of the sediment is deposited upstream of the high water level and, as indicated further on, affects upstream flooding. As shown in figure 3.2, as the delta gradually moves into the reservoir, it affects the active storage space very early on.[6]

These are very important observations to be aware of when investigating the feasibility of dam and reservoir projects. An old and common—but incorrect—assumption is that dead storage space is reserved for deposited sediment. This is clearly not the case. Sediment deposition in the active storage space is as prevalent and common as in the dead storage space, particularly in large reservoirs. It is necessary to account for such loss in active reservoir storage space early on in the life of a project and to recognize its impact on the reliability of water and power supply, and on flood control.

Hydropower

Sedimentation at hydropower plants affects two aspects of hydropower production: the amount of power produced and maintenance requirements for turbines. Power production is limited when the active reservoir storage is diminished as a result of sedimentation. Maintenance requirements increase if the sediment flowing through the turbines contains high levels of hard minerals, causing severe abrasion of turbine parts.

Hydropower Generation

Okumura and Sumi (2013) investigate the impact of reservoir sedimentation on hydropower generation in Japan. They find that the water use efficiency of a hydropower plant steadily decreases as sedimentation in the active storage volume increases. Water use efficiency is expressed as the total volume of water discharged through the turbines divided by the total volume of water flowing into the reservoir. Figure 3.3 shows that water use efficiency steadily decreased over four decades as sedimentation increased, eventually occupying about 23 percent of the original active storage volume.

Figure 3.4 shows the performance of five other hydropower plants reviewed by Okumura and Sumi (2013). Four of those plants are storage hydropower projects (A through D) and one is a run-of-river project (F). Each of the points on the graph represents the average water use efficiency for a decade, plotted against the percentage sedimentation of the active storage space. Of course, sedimentation is not the only factor determining water use efficiency. The skill of the operators also plays a role, leading to the positive trend at some of the plants in spite of sedimentation. It can nevertheless be concluded that sedimentation of the active storage space has a detrimental impact on water use efficiency. This negative impact is observed in both storage and run-of-river facilities. Rapid assessment of the impacts of storage loss on power generation can be accomplished by using the methods developed by Xie, Annandale, and Wu (2010) and Xie, Wu, and Annandale (2013).

Extending the Life of Reservoirs • http://dx.doi.org/10.1596/978-1-4648-0838-8

Figure 3.3 Changes in Water Use Efficiency Relative to Sedimentation in the Active Storage of a Reservoir over Four Decades

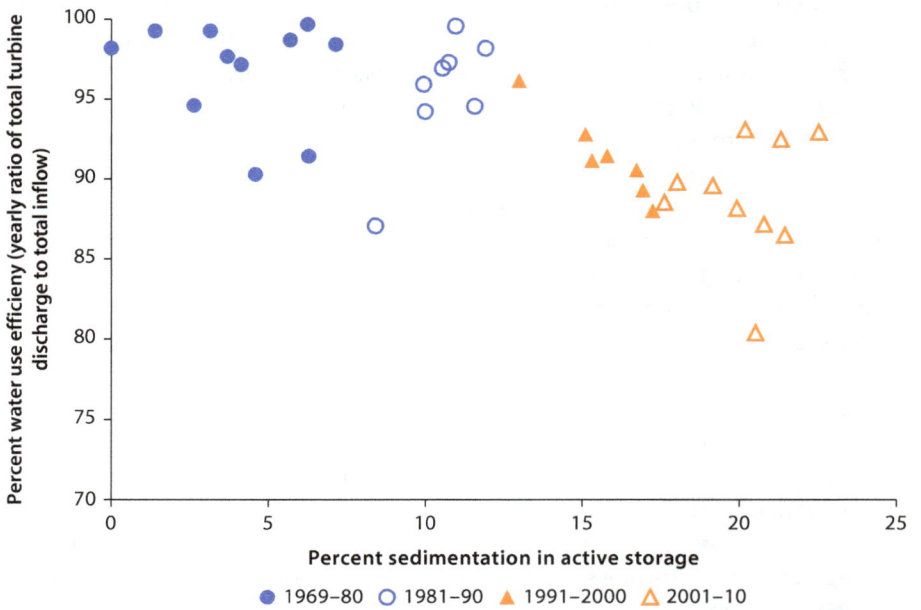

Source: Adapted from Okumura and Sumi 2013.

Figure 3.4 Changes in Water Use Efficiency Relative to Sedimentation in the Active Storages of Reservoirs

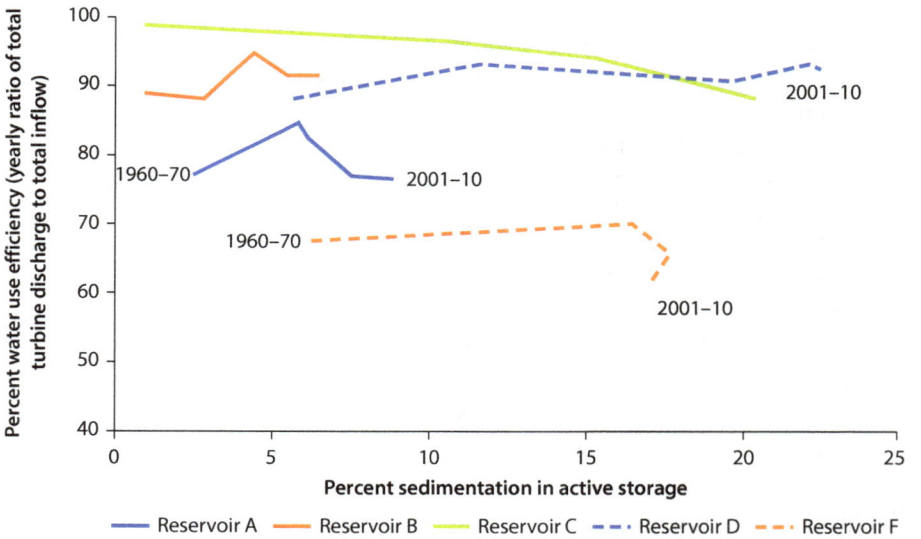

Source: Adapted from Okumura and Sumi 2013.
Note: Figure shows changes in water use efficiency at five hydropower plants relative to sedimentation in the active storage of a reservoir. Reservoirs A through D are storage hydropower plants; F is a run-of-river hydropower plant.

Abrasion

Abrasion occurs when the sediment contained in the water flowing through turbines contains minerals that are harder than the metal used to manufacture the turbine. Minerals of interest are quartz, feldspar, tourmaline, and other minerals with Mohs hardness greater than 5. When water containing sediment flows through a turbine, it can result in abrasion of the wet parts, for example, the runners and wicket gates.

Plants are often designed to remove most of the coarse sediment particles, while allowing fines (silt and clay) to flow through the turbines. However, even silt can cause significant abrasion if the quartz content of the silt and the pressure head are high enough. For example, the 1,500 megawatt Nathpa Jhakri hydropower plant in India contains four desilting chambers that are 525 meters long, 16.3 meters wide, and 27.5 meters high. The intent of these chambers is to remove coarse sediment before the water flows into the turbines (photo 3.1). In spite of the removal of the coarse sediment, the fines remaining in the water discharging through the turbines at Nathpa Jhakri caused significant abrasion. Not only is the quartz content of the silt very high, but the pressure head at the turbines is also high at 428 meters. The abrasion at Nathpa Jhakri after commissioning (photo 3.2) was so severe that wet parts of the turbines had to be replaced within one year.

Water Supply

Reservoir storage, as already indicated, is critically important to ensuring reliable supply of water. A reservoir's required volume depends on the average

Photo 3.1 Dewatered Desilting Chambers at Nathpa Jhakri Hydropower Plant, 2010

Source: © Nathpa Jhakri. Used with permission. Further permission required for reuse.

Photo 3.2 Abrasion of Wicket Gates at Nathpa Jhakri Plant after Five Months of Operation

Source: © Nathpa Jhakri. Used with permission. Further permission required for reuse.

annual flow in the river, its hydrologic variability, the demand for water, and the required reliability of water supply. If reservoir storage volume is reduced by reservoir sedimentation, the amount of water that can be reliably supplied will also be reduced.

Figure 3.5 relates dimensionless yield and the annual coefficient of variation of stream flow for three dimensionless reservoir volumes: 1 MAF, 0.5 MAF, and 0.25 MAF. The volume of water in a reservoir can be expressed in a dimensionless manner by dividing the reservoir volume by the average volume of water that flows in the river entering it (MAF). Yield can similarly be expressed in dimensionless terms by dividing the amount of water a reservoir can yield at a certain reliability by the MAF.

Figure 3.5 illustrates the combined effect of reservoir sedimentation and climate change. Assume, for example, that a particular dam was originally built with a reservoir volume equaling 1 MAF and that the annual coefficient of variation when the dam was built equaled 0.4 (point A). From the figure it can be seen that the amount of water that could be reliably yielded at 99 percent reliability is about 0.92. Should climate change lead to an increase in hydrologic variability from 0.4 to 0.6, say, then the yield would decline to about 0.88 (point B). This is a fairly small reduction in yield.

However, if the reservoir volume declines during the same period that the effects of climate change are setting in, the yield may drop significantly more.

Figure 3.5 Relationship between Dimensionless Yield and Dimensionless Reservoir Storage for Varying Hydrologic Variability

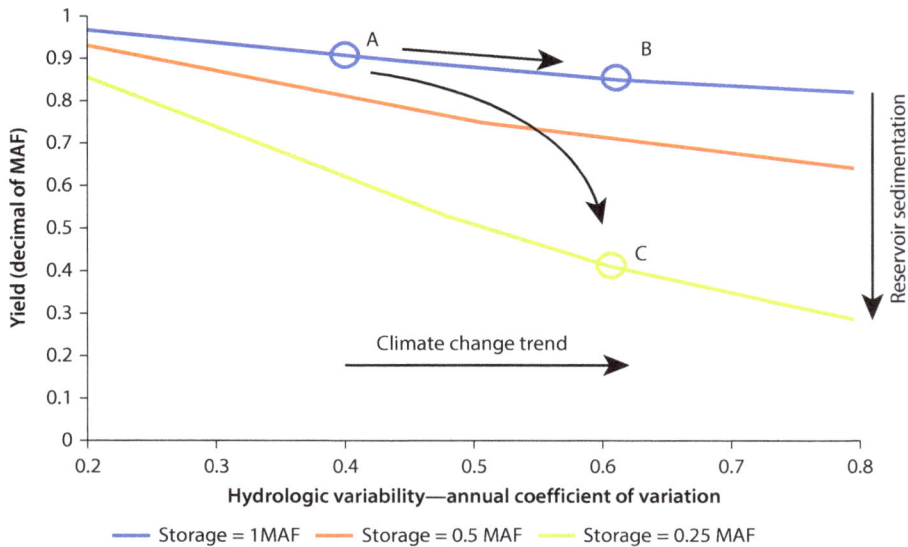

Source: Based on McMahon et al. 2007.
Note: MAF = mean annual flow. Figure shows relationship between dimensionless yield and dimensionless reservoir storage for reliability of supply equal to 99 percent.

For example, if the reservoir volume decreases to, say, 0.25 MAF because of reservoir sedimentation, and climate change concurrently leads to an increase in the annual coefficient of variation equaling 0.6, then the yield declines to about 0.40 (point C). The combined effect of reservoir sedimentation and climate change leads to a significant reduction in yield, from about 0.92 to 0.40.

Clearly, the effects of both climate change and storage loss to reservoir sedimentation require serious consideration when evaluating dam and reservoir projects, particularly when taking account of the potential effects on future generations.

As shown in panel a of figure 3.6, as reservoir sedimentation progresses, reservoir volume diminishes, leading to a reduction in water supply. Initially, the decrease in water supply is very small and almost indiscernible. As reservoir sedimentation progresses, however, water supply eventually drops rapidly. Once that happens, it is often too late to do anything about it. Removing sediment filling a large reservoir is very difficult and costly. Restoring the water supply function from such a reservoir becomes very difficult, if not impossible.

A better approach would be to regularly remove deposited sediment from a reservoir, as illustrated in panel b of figure 3.6. By doing so, the problem becomes manageable and the water supply from the reservoir is maintained over the long term. Various methods for sediment management are presented starting in chapter 7.

Figure 3.6 Positive Effect on Water Supply of Reservoir Sediment Management

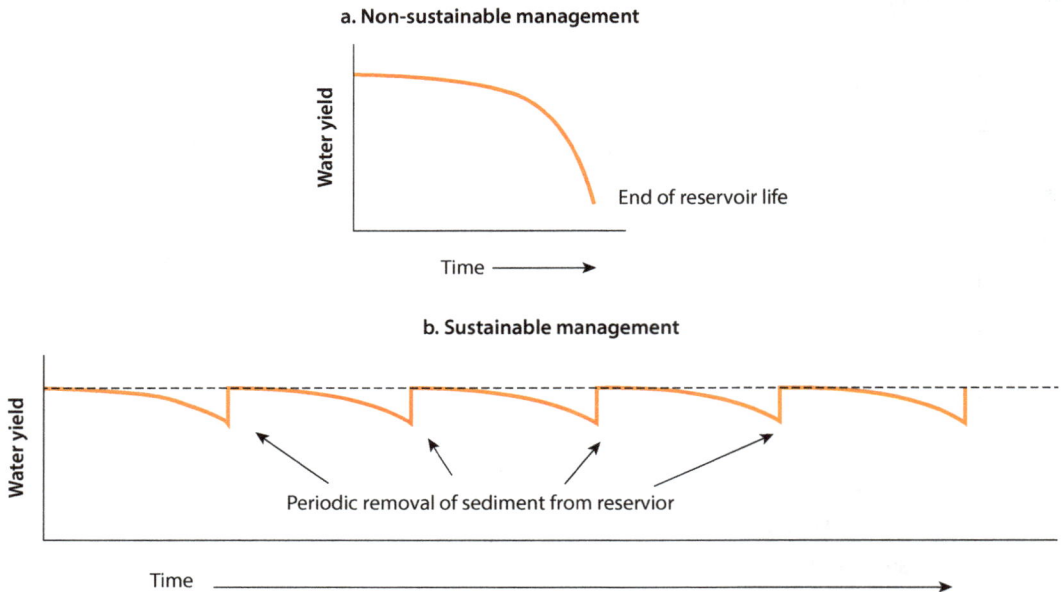

a. Non-sustainable management

b. Sustainable management

Increased Flooding

As illustrated in figure 3.2, some of the sediment deposited upstream of a reservoir is located above the high water level. This deposition occurs because backwater effects cause water flowing into the reservoir to slow down upstream of the high water level. The sediment that deposits upstream of a reservoir reduces the water-carrying capacity of the river channel and results in increased flood levels upstream of the reservoir (figure 3.7).

Recreation, Environment, and Other Impacts

Sedimentation upstream of dams affects the environment as well as recreation opportunities and property values. Rivers and dams are unique, and impacts at each should be identified as distinctive features.

An interesting example of the impact of sedimentation upstream of dams is found in the Lower Mekong River flowing through the Lao People's Democratic Republic, Cambodia, and Vietnam in Southeast Asia. This river contains very deep pools, some up to 100 meters deep, providing resting places for aquatic creatures like the Irrawaddy dolphin and other fish species. The planned construction of a number of dams in this river may result in sedimentation of the pools located upstream of the dams, which will obviously impact biodiversity

An example of reservoir sedimentation affecting property values can be found along the shores of Lewis and Clark Lake upstream of Gavin's Point Dam on the Missouri River, in the United States (photo 3.3).

Extending the Life of Reservoirs • http://dx.doi.org/10.1596/978-1-4648-0838-8

Severe sedimentation of the lake upstream of this dam resulted in phragmites becoming established on the sediment, which severely reduced property values along the shoreline because of the impact on scenic beauty. Establishment of this vegetation also compromises efforts to remove sediment from the reservoir. The phragmites bind the sediment and make it difficult to remove.

Figure 3.7 Increased Flood Elevations Caused by Sediment Deposition

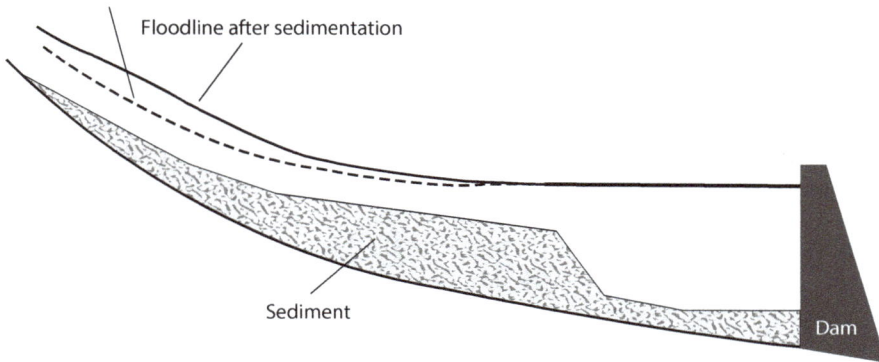

Photo 3.3 Phragmites Established on Deposited Sediment in Lewis and Clark Lake

Source: © George W. Annandale. Used with permission. Further permission required for reuse.

Sedimentation upstream of reservoirs may also affect the ability of bridges to pass flowing water. As the size of the bridge opening diminishes, the amount of water that can be passed by such a bridge will also decrease. An example of such a case is found at Welbedacht Dam in South Africa where the Jim Fouche Bridge had to be replaced because of the effects of reservoir sedimentation upstream of the dam, beyond the full supply level (de Villiers and Basson 2007).

Sedimentation Impacts Downstream of a Dam

Fluvial Morphology

When reservoirs capture sediment, the amount of sediment in the water released downstream of the dam is reduced, compared with its historic presence. The water downstream of the dam is sometimes referred to as "sediment hungry" water. The implication is that the water flowing in the downstream river has greater capacity to carry additional sediment. The net effect is that the river erodes and degrades (figure 3.8).

Aquatic Ecosystems

Fine sediments such as silt and clay carry nutrients required to produce food consumed by fish. When such sediments are captured in reservoirs upstream of dams, fewer nutrients are released downstream. Decreases in nutrients affect fishery populations and the aquatic ecosystem. In addition to the impact on food availability for fish, the presence of sediment hungry water results in degradation of aquatic habitat.

Coastal Impacts

Beach sand along coast lines consists principally of sediment discharged into oceans by rivers. If the amount of sediment discharging into oceans from rivers is reduced because of reservoir sedimentation, beaches can deteriorate.

Figure 3.8 Erosion and Degradation of Downstream Rivers Due to "Sediment Hungry" Water

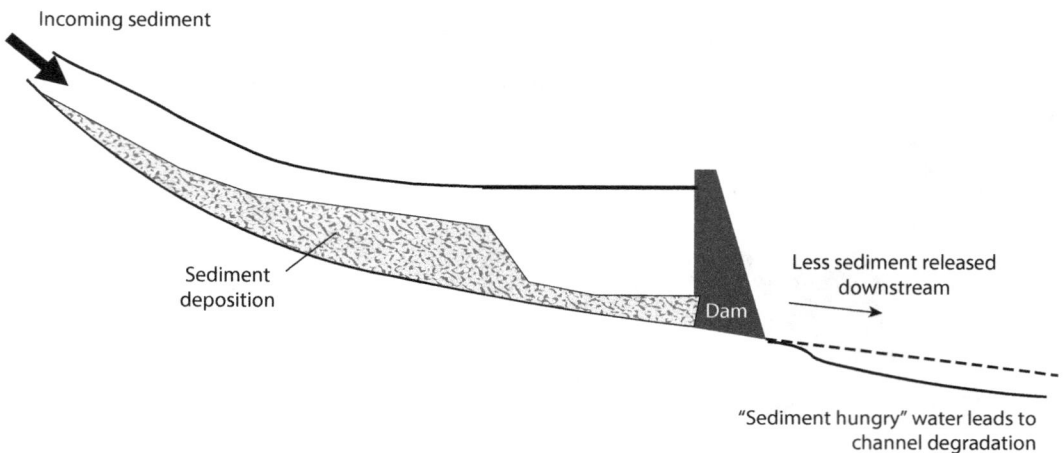

Incoming sediment

Sediment deposition

Dam

Less sediment released downstream

"Sediment hungry" water leads to channel degradation

An example illustrating the impact of reservoir sedimentation on beach erosion is shown in photo 3.4. The photo contains three aerial shots of the outfall of the Tenryu River in Japan taken in 1946, 1961, and 2001. The dams that were constructed at different times are listed along the left side of the photographs. The net effect of reducing the amount of sediment transported by the Tenryu River can be observed. The extent of the beach in 2001 is significantly smaller than it was in 1946. Japan is currently developing an extensive project to pass sediment through the reservoirs along the Tenryu River and to introduce sediment into the river downstream of the reservoirs, with the intent of restoring the beach over the long term.

Flood Management

The use of reservoirs for effective flood management requires enough reservoir storage space to temporarily store flood waters for gradual release downstream. This operation is known as flood attenuation. The degree to which a flood can be attenuated is determined by the amount of reservoir storage available, and by the operating procedure at the reservoir.

The contribution of flood control storage to flood control benefit is shown in figure 3.9. The figure shows that the annual average flood control benefit for Three Gorges Dam increases with increasing flood control storage volume. If reservoir sedimentation decreases the flood control storage volume, flood control benefits will be decreased.

Photo 3.4 Impact of Reservoir Sedimentation at the Mouth of the Tenryu River, Japan

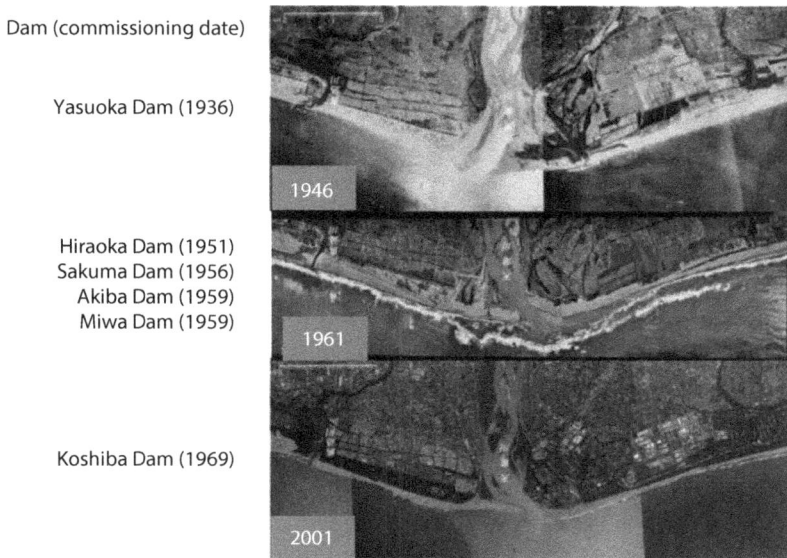

Source: Sumi 2003. © T. Sumi. Used with permission. Further permission required for reuse.

Extending the Life of Reservoirs • http://dx.doi.org/10.1596/978-1-4648-0838-8

Figure 3.9 Relationship between Annual Average Flood Control Benefit and Flood Control Storage for Three Gorges Dam, China

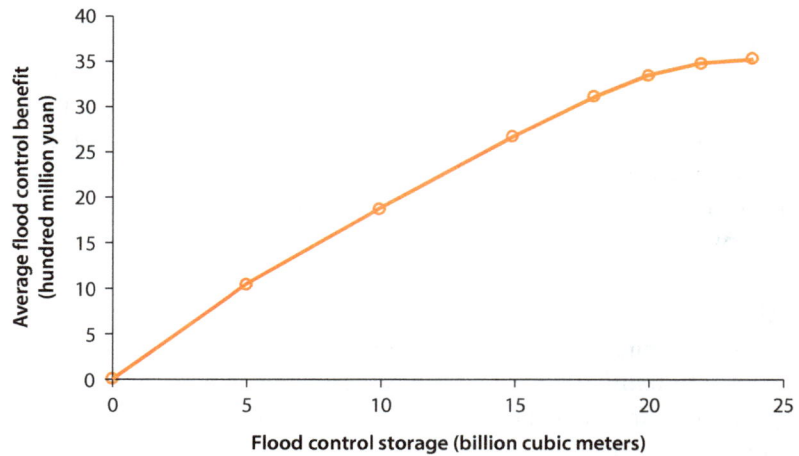

Source: Xie, Wu, and Annandale 2013.

Importance of Sediment Management

Reservoir sediment management is vitally important for preserving reservoir storage space and minimizing long-term maintenance costs. Sediment management techniques that may be used to accomplish this goal are discussed starting in chapter 7.

For purposes of sustainable development, classifying reservoir storage space as either an exhaustible or a renewable resource depends on choices made by designers and operators of dams and reservoirs. If a conscious decision is made to allow a reservoir to fill with sediment, it is deliberately classified as an exhaustible resource. If, however, the design and operation of a dam and reservoir focus on preserving reservoir storage space, it is classified as a renewable resource, by choice. Historically, designers and operators of dams and reservoirs assumed that storage loss to reservoir sedimentation was inevitable, effectively categorizing those reservoirs as exhaustible resources.

To illustrate the point, consider a real-life situation. Figure 3.10 shows the cumulative yield as a function of unit cost for all potential dam and reservoir sites in Kenya. The horizontal axis of the figure represents the construction cost per cubic meter of the amount of water that a reservoir may yield on an annual basis. The vertical axis represents the cumulative yield from all identified reservoir sites in Kenya, should they be built. Cumulative yield is determined by summing the annual yield from each reservoir when they are ranked from the lowest to the highest unit cost.

Assuming that those dams and their reservoirs would be developed starting with the least-cost reservoirs, the impact on sustainable development of prevailing engineering design philosophy is shown in figure 3.11. Contemporary engineering design philosophy generally assumes that it is acceptable for a reservoir to fill with sediment over a period known as its "design life." If a dam and its reservoir are shown to be economically viable, the current assumption is that it does not matter if a reservoir completely fills with sediment. This approach obviously does not contribute to long-term infrastructure sustainability.

Figure 3.11 contains four panels, which should be viewed from the top left to top right, and then from the bottom left to the bottom right. Note that the top left panel is similar to figure 3.10, that is, it represents all potential dam and reservoir sites that might be developed in Kenya.

Assume that the initial demand for water[7] that must be satisfied by the first generation of dam builders is 20 billion cubic meters per year (shown by the horizontal dashed line). If that demand can be satisfied by building the most cost-effective dams first, it is only necessary to build a few dams at very low cost.

By accepting current design philosophy, that is, allowing the reservoirs to fill with sediment over their design lives, the number and quality of the remaining potential dam and reservoir sites declines once the first set of reservoirs are all filled with sediment. The remaining dams and reservoirs are shown in the top right panel of figure 3.11. These are the dam and reservoir sites available to the second generation of dam builders.

At that time, assume that the demand for water may have increased to, say, 30 billion cubic meters per year. The panel shows that many more dams

Figure 3.10 Cumulative Yield as a Function of Unit Cost for All Potential Dam and Reservoir Sites in Kenya

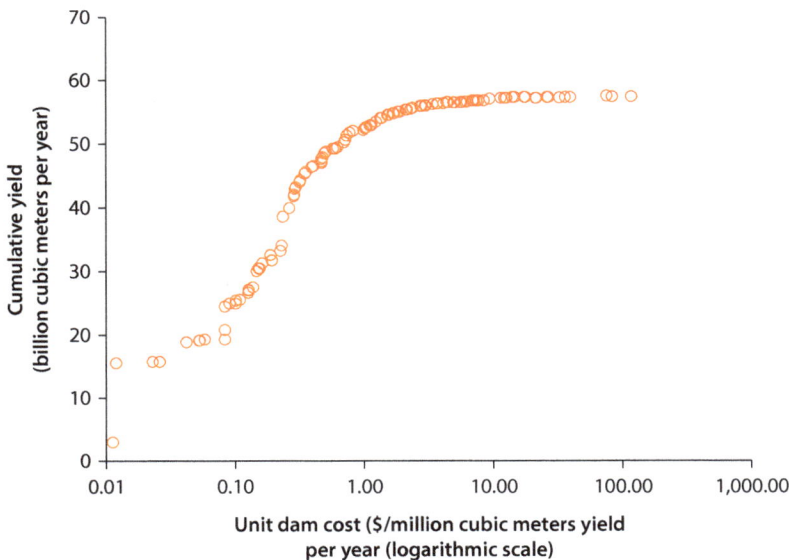

Source: Adapted from Annandale 2013.

Extending the Life of Reservoirs • http://dx.doi.org/10.1596/978-1-4648-0838-8

Figure 3.11 Adverse Effect of Developing Dams and Their Reservoirs in a Nonsustainable Manner

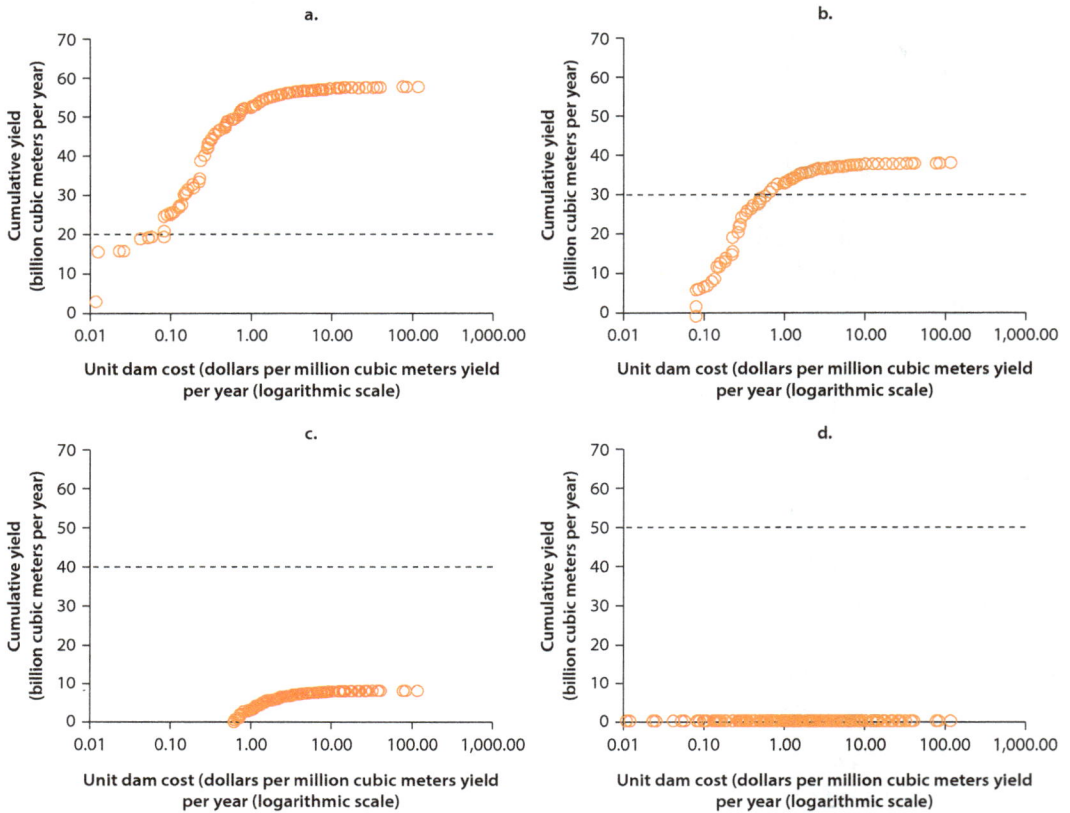

Source: Adapted from Annandale 2103.

at much higher cost should be constructed to satisfy the increased demand for water. Fortunately, at that point, sufficient dam and reservoir sites are still available for construction.

Using the same design philosophy, all of the second set of reservoirs will be filled with sediment at some future point. At that time, the demand for water has increased to, say, 40 billion cubic meters per year, and the third generation of dam builders has a problem. The bottom left panel of figure 3.11 shows that the demand for water far exceeds the ability of the remaining number of dams and reservoirs to satisfy this need. In addition, dam construction costs are very high.

If the third generation of dam builders constructs dams and reservoirs on all these sites and allows them to fill with sediment, a severe problem occurs. While the demand for water for the fourth generation of dam builders might have risen to, say, 50 billion cubic meters per year, no more dam and reservoir sites exist, as shown in the bottom right panel of figure 3.11.

The lesson is that the decision—the choice—to allow reservoirs to fill with sediment characterizes the reservoir storage space as an exhaustible resource.

The fact that current design norms deem it acceptable for reservoirs to fill with sediment over a set period determines the nature of the reservoir storage space: it is classified as exhaustible through a deliberate decision by the developer.

Conversely, if a decision were made to prevent or minimize storage loss by managing reservoir sedimentation, the classification of the storage space changes from exhaustible to renewable. If storage loss from reservoir sedimentation could be entirely prevented through implementation of reservoir sediment management approaches, reservoirs could be classified as renewable resources. In such a case, a reservoir could potentially be used in perpetuity, satisfying the tenets of sustainable development.

The decision to use reservoir storage space either as a renewable or an exhaustible resource is made very early on during the project development phase. To accomplish sustainable development goals, the economic analysis of dams and their reservoirs must be correctly executed. The shortcomings of current approaches to the economic analysis of dams are dealt with in chapter 2.

A reservoir that is allowed to completely fill with sediment eventually reaches a new morphologic equilibrium once the reservoir storage is reduced to zero. Figure 3.12 shows the change in reservoir storage volume over time. Note that at some point the reservoir storage volume is reduced to zero. The broad arrow in the figure indicates the total volume of sediment that remains in such a reservoir over the long term. It is equal to the total volume of sediment that has been prevented from entering the river downstream of the dam, resulting in downstream river and aquatic habitat degradation.

Figure 3.12 Long-Term Reduction in Reservoir Storage Space from Reservoir Sedimentation

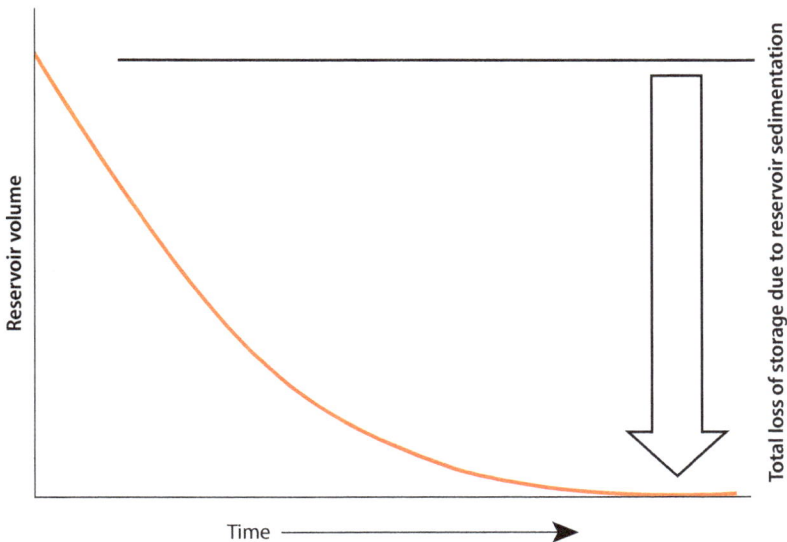

Figure 3.13 The Effect of Successful Reservoir Sediment Management

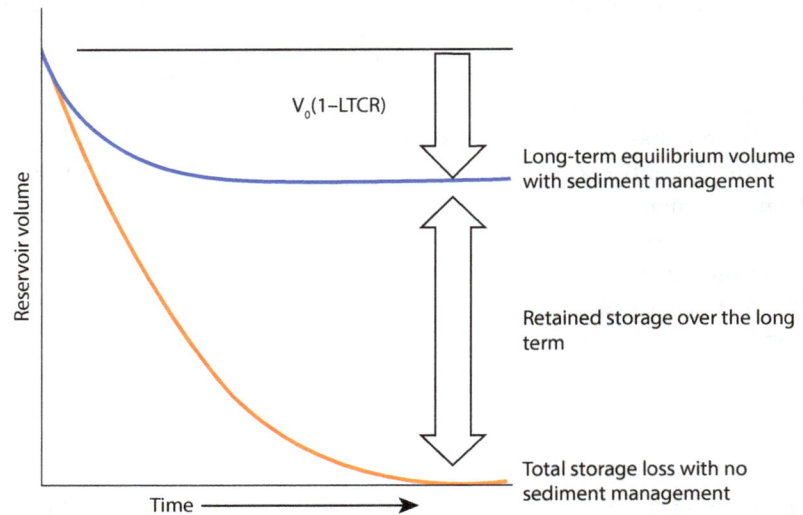

Note: LTCR = long-term capacity ratio.

The objective of reservoir sediment management is to increase the amount of storage that will be available in a reservoir over the very long term, that is, to create a new equilibrium characterized by a larger remaining reservoir volume. That new equilibrium volume, when established by means of regular reservoir sediment management, is represented by the long-term capacity ratio (LTCR). The LTCR is the percentage of the original storage that remains once the new equilibrium resulting from reservoir sediment management has been reached. For example, if the LTCR is, say, 80 percent, then 80 percent of the original reservoir volume (V_0) can be retained indefinitely in the long term through the use of reservoir sediment management.

The effect of successful reservoir sediment management is shown in figure 3.13. The figure shows that the remaining long-term reservoir storage volume is larger with sediment management than it is with no sediment management. The total amount of sediment retained is equal to $V_0 \times (1-LTCR)$, which is less than the amount of sediment that would have been deposited otherwise. This new equilibrium results in a win-win situation. Reservoir storage is preserved while downstream river and aquatic habitat degradation are concurrently reduced, and the amount of nutrients discharged downstream is increased.

Severity of Storage Loss to Sedimentation

The global net amount of reservoir storage space has been decreasing in recent years because reservoir sediment management was not standard practice in the past. This trend is the result of a decrease in the rate at which reservoir storage

Extending the Life of Reservoirs • http://dx.doi.org/10.1596/978-1-4648-0838-8

Figure 3.14 Global Population Growth and Reservoir Storage Volume

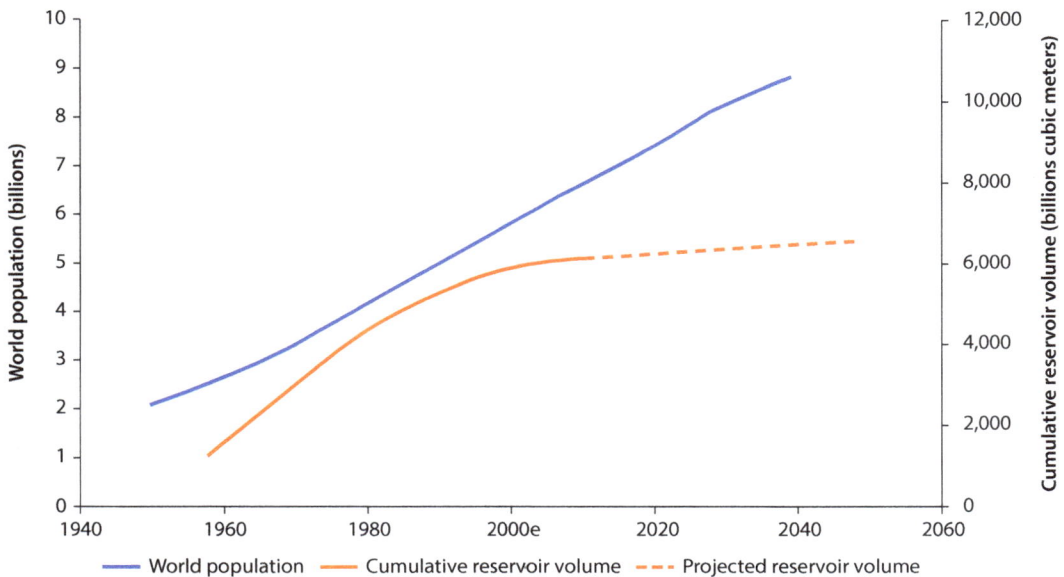

Source: Annandale 2013.

has been added since about 1980 (figure 3.14) and the continued loss of storage space to reservoir sedimentation (figure 3.15). Figure 3.15 shows that total net reservoir storage space, after accounting for storage loss due to sedimentation, has decreased since about 2000, while global storage space per capita has decreased since about 1980. The current per capita net reservoir storage space roughly equals what it was in 1965. The importance of implementing reservoir sediment management techniques to preserve reservoir storage space is evident.

Sedimentation and Climate Change

Climate change can lead to increased sediment loads in rivers, resulting in increased amounts of sediment deposition in reservoirs. The exact impact of climate change on the amount of sediment carried by rivers is not known, although studies indicate that increases in sediment yield are more likely than decreases (for example, Shrestha et al. 2013), which could result in increased amounts of sediment depositing in reservoirs.

The impact of storage loss from reservoir sedimentation on water and power supply reliability, and on flood control efficiency, will be more severe under climate change conditions. Increased hydrologic variability will require larger reservoir storage volumes to maintain these functions (see "The Importance of Storage" section in this chapter and figure 3.5).

Figure 3.15 Net Global Reservoir Storage Volume, Accounting for Storage Loss from Reservoir Sedimentation

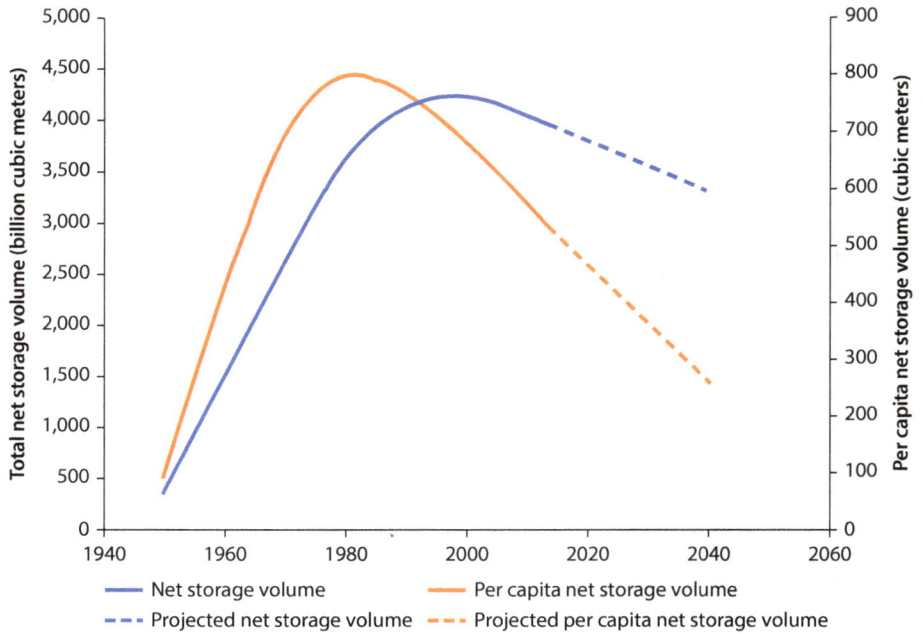

Source: Annandale 2013.

Notes

1. Hydrologic variability is expressed by the annual coefficient of variation of river flow, which is the standard deviation of annual river flow volumes divided by the mean annual river flow volume.

2. This curve is based on a standardized probability equation for dimensionless yield from a river without the presence of a dam, $\alpha = 1 + C_v \times z_p$, where α = dimensionless yield, C_v = annual coefficient of variation of flow, and z_p = standardized log-normal distribution deviate.

3. A yield value of 50 percent, for example, means that the amount of water that may be reliably supplied equals 50 percent of the mean annual flow volume in the river.

4. Note that the graph illustrates general trends by assuming a log-normal probability distribution for river flows. Proposed projects will require site-specific analyses.

5. Chapter 4 presents other distribution patterns of deposited sediment, which the reader may use in conjunction with the information in this chapter to interpret what may happen in other cases.

6. As the reader may discover in chapter 4, the impact of other distribution patterns may differ. For example, if density currents dominate, sediment deposition would be concentrated close to the dam, and a more gradual deposition would also occur upstream.

However, when coarser sediments (coarse sand and gravel) are mostly present, delta formation is common in large reservoirs.

7. Please note that the assumption on water demand is fictitious. It is used to illustrate the potential exhaustible nature of reservoir storage space.

References

Annandale, G. W. 2013. *Quenching the Thirst: Sustainable Water Supply and Climate Change*. Charleston, SC: CreateSpace.

———. 2015. "Policy Considerations for Sustainable Hydropower—Reliability, Climate Change, and Sedimentation." Presentation at HYDRO 2015, Bordeaux, France, October 26–28.

De Villiers, J. W. L., and G. R. Basson. 2007. "Modeling of Long-Term Sedimentation at Welbedacht Reservoir, South Africa." *Journal of the South African Institution of Civil Engineering* 49 (4): 10–18.

IPCC (Intergovernmental Panel on Climate Change). 2013. *Climate Change 2013: The Physical Science Basis. Working Group I Contribution to the Intergovernmental Panel on Climate Change Assessment Report Number 5*, edited by T. F. Stocker, D. Qin, G.-K. Plattner, M. Tignor, S. K. Allen, J. Boschung, A. Nauels, Y. Xia, V. Bex, and P. Midgley. New York: Cambridge University Press.

Okumura, H., and T. Sumi. 2013. "Reservoir Sedimentation Management in Hydropower Plant Regarding Flood Risk and Loss of Power Generation." Proceedings of the International Symposium on Dams for a Changing World, The 80th Annual Meeting of ICOLD.

McMahon, T. A., G. G. S. Pegram, R. M. Vogel, and M. C. Peel. 2007. "Review of Gould-Dincer Reservoir Storage-Yield-Reliability Estimates." *Advances in Water Resources* 30: 1873–82.

Rahman, K., S. M. Gorelick, P. J. Dennedy-Frank, and J. Yoon. 2015. "Declining Rainfall and Regional Variability Changes in Jordan." *Water Resources Research* 51: 3828–35.

Shrestha, B., M. S. Babel, S. Maskey, A. van Griensven, S. Uhlenbrook, A. Green, and I. Akkharath. 2013. "Impact of Climate Change on Sediment Yield in the Mekong River Basin: A Case Study of the Nam Ou Basin, Lao PDR." *Hydrology and Earth System Sciences* 17: 1–20.

Sumi, T. 2003. "Approaches to Reservoir Sedimentation Management in Japan." *Reservoir Sedimentation Management Symposium*, Third World Water Forum, Kyoto, Japan.

Xie, J., B. Wu, and G. W. Annandale. 2013. "Rapid Reservoir Storage–Based Benefit Calculations." *Journal of Water Resources Planning and Management* 139 (6): 712–22.

Xie, J., Annandale, G.W. and Wu, B. 2010. Reservoir Capacity-Potential Power Generation-Reliability Estimation Based on Gould-Dincer Approach, Proceedings of the 34th World Congress of the International Association for Hydro-Environment Research and Engineering, Australia.

CHAPTER 4

Sediment Yield

George W. Annandale

Introduction

Estimates of the amount of sediment transported by rivers are important for evaluating the impacts of reservoir sedimentation and how to manage it. This chapter provides an overview of global sediment yield, how it relates to catchment conditions, and its spatial and temporal variation. Modes of sediment transport are briefly presented, followed by ways to measure sediment yield. Methods for estimating sediment yield, including those that can be used when sediment data are not available, are discussed.

Global Sediment Yield: Spatial Variability

Sediment yield is the amount of sediment carried by rivers. The total sediment load in rivers is the sum of wash load, suspended load, and bed load. Bed load consists of coarse sediment particles located on a river bed and dragged along by flowing water. It moves along the bed of a river by rolling, sliding, and saltating (hopping). If the sediment transport capacity of the flowing water increases, the amount of bed load transport increases as well. At some point, the bed material that saltates may become suspended in the water column; this is known as "suspended load." Suspended load normally consists of finer sediment particles that are light enough for the turbulence in the water to retain them in suspension. Obviously, if the sediment transport capacity of the flowing water decreases, thereby decreasing the turbulence in the water, some of the sediment particles carried in suspension may deposit back onto the riverbed and become bed load again.

Wash load usually consists of very fine particles, such as clay particles or silt. Such particles are easily suspended in the water column and may sometimes remain in suspension even if the sediment transport capacity decreases to very low values. Positive and negative electric charges on these fine particles cause

Extending the Life of Reservoirs · http://dx.doi.org/10.1596/978-1-4648-0838-8

them to perform an elegant dance in the water, even in stagnant water, thereby remaining in suspension.

It is important to note that, as far as reservoir sedimentation is concerned, 100 percent of bed load is normally trapped by reservoirs. A portion of or all of the suspended load carried by a river into a reservoir may be deposited within the reservoir. Depending on the hydraulic characteristics of flow through a reservoir and on the sizes of suspended sediment carried by the flow, some of the suspended sediment may flow through a reservoir and over the dam for release downstream.

Sediment yield estimates mostly consist of the amount of suspended sediment carried by rivers because estimates of bed load carried by rivers are generally unavailable. If present, wash load may be included in the estimates of suspended sediment load, but it is generally not reported separately.

Two terms are used to identify the amount of sediment flowing in rivers— sediment yield and specific sediment yield. Sediment yield is usually expressed in tons of sediment per year. Specific sediment yield is usually expressed as tons per unit catchment area (in square kilometers, or km^2) per year ($t/km^2/yr$).

Sediment yield is customarily expressed in terms of mass (that is, tons, kilograms, and the like). Expressing it as volume (cubic meters, cubic feet, and so on) is less desirable. The bulk density of sediment may vary, making it difficult to provide consistent estimates of the volume of sediment discharging in a river.

Sediment yield varies globally, depending on climate, lithology, topography, human-influenced soil erosion, forest fires, catchment area, river discharge, temperature, and the trap efficiency of upstream reservoirs. An indication of global differences in sediment yield is provided in map 4.1, which shows specific

Map 4.1 Global Specific Sediment Yield Map

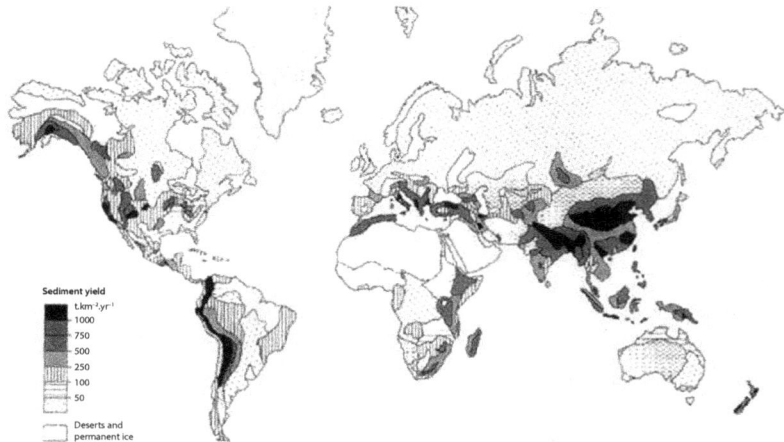

Source: Walling and Webb 1983. Used with permission of the author. Further permission required for reuse.
Note: t.km^{-2}. yr^{-1} = tons per square kilometer per year.

sediment yield in metric t/km²/yr, and in table 4.1. Map 4.1 shows that specific sediment yield can be very low in some regions (50 t/km²/yr and lower) and very high in others (1,000 t/km²/yr and higher).

Table 4.1 presents an estimate of the total amount of suspended sediment discharging to the oceans from the continents. Note that the highest specific sediment yields originate from Oceania and the Pacific Islands and from Asia. South America also has a high specific sediment yield, with Africa and Europe accounting for the lowest sediment yields. However, large parts of the African land mass consist of deserts, where the sediment yield is very low. If the combined land area of the Sahara, Namib, and Kalahari Deserts (about 10.4 million square kilometers) is subtracted from the total land area of Africa, the sediment yield from the remaining land mass is estimated to be about 108 t/km²/yr. Table 4.2 shows sediment yield for 10 large rivers, ranging from 160 million tons per year to 1,670 million tons per year.

Table 4.1 Sediment Yield from the Continents to the Oceans

Continent	Land area (million km²)	Mean annual runoff (thousand km³)	Total annual suspended sediment load (million tons/year)	Average specific sediment yield tons/km²/year
Africa	15.3	3.4	530	35 (108)[a]
Asia	28.1	12.2	6,433	229
Europe	4.6	2.8	230	50
North and Central America	17.5	7.8	1,462	84
Oceania and Pacific Islands	5.2	2.4	3,062	588
South America	17.9	11.0	1,788	100

Sources: Martin and Meybeck 1979; Milliman and Meade 1983; and Walling 1987.
Note: km = kilometers.
a. Sediment yield from nondesert land area in Africa.

Table 4.2 Average Sediment Discharge (Yield) for 10 Large Rivers

River and country	Average sediment discharge (million tons/year)
1. Ganges-Brahmaputra, India	1,670
2. Yellow, China	1,080
3. Amazon, Brazil	900
4. Yangtze, China	478
5. Irrawaddy, Myanmar	285
6. Magdalena, Colombia	220
7. Mississippi, United States	210
8. Orinoco, Venezuela, RB	210
9. Hungho (Red), Vietnam	160
10. Mekong, Thailand	160

Source: Milliman and Meade 1983.

Sediment Sources

Weathering and Hill Slope Erosion

Weathering of bedrock from both physical and chemical processes generates sediment on hill slopes. Physical processes may include movement in rock caused by temperature variations and expansion of ice that may form in crevices and discontinuities in the rock. Such expansion may lead to breakage of the rock. Similarly, erosion from wind and rain may occur over time. Sediment that has moved downhill to the bottom of the slope without the help of running water in streams (that is, gravity caused it to move) is known as colluvium.

The force of rain falling on the colluvium loosens sediment particles, and runoff generated by the rainfall transports the sediment into channel networks. The overland flow first erodes the sediment as sheet flow, which turns into rills and gullies that increase the amount of erosion and material transported to the channel network (photo 4.1).

Rivers located in areas with fragile geology also receive sediment from landslides, debris flows, and mud flows. These sediment sources are prevalent in the Himalayas, Andes, and areas characterized by seismic activity, including Central America, Indonesia, Japan, New Zealand, and other countries.

Catchment Conditions and Sediment Yield

Important Factors

Important natural factors determining the magnitude of sediment yield are geology, topography, and climate, while human influence can significantly exacerbate erosion and sediment yield. Geologic factors of note are the percentage of exposed bedrock and soil characteristics. For example, clay may be more difficult

Photo 4.1 Erosion: Sheet Flow, Rill Erosion, Gully Erosion

Source: © Golder Associates, Perth. Used with permission. Further permission required for reuse.

to erode than sandy material, and gravel more difficult than sand, with bedrock taking much longer to erode than any of these.

Important topographic features are steepness of terrain and drainage density. Water flowing down steep slopes has greater potential to cause erosion than water flowing along mild slopes. Regions with very high drainage density, that is, a large number of connecting streams and channels in close proximity to one another, are characteristic of high sediment yield.

The impact of climate on sediment yield is illustrated in panel a of figure 4.1, which indicates that sediment yield from desert shrub areas can vary significantly from low to very high, whereas yield from grassland areas can range from high to medium, and forest areas are characterized by low sediment yield.

Panel b of figure 4.1 illustrates the influence of mean annual rainfall and geology on sediment yield. It indicates that for Rajasthan, India, sediment yield increases with mean annual rainfall and that the yield from a limestone region is higher than that from a sandy region, which in turn is higher than that from alluvium. Note that the sediment yield from a particular soil type can easily range over two orders of magnitude, depending on the mean annual rainfall.

These large ranges of sediment yield from a particular geology have also been measured in the United States, as reported by Vanoni (1975). Many of the measurements plotted in figure 4.2 range up to two to even three orders of magnitude for a particular geology type, similar to the ranges shown in panel b of figure 4.1. Clearly, estimating sediment yield by merely accounting for geologic type is not possible; the other factors need to be incorporated as well.

Estimating sediment yield is not an exact science and is made more difficult by the paucity of relevant field data. The best way to estimate sediment yield

Figure 4.1 Specific Sediment Yield as a Function of Effective Precipitation and Terrain and as a Function of Mean Annual Precipitation and Geology

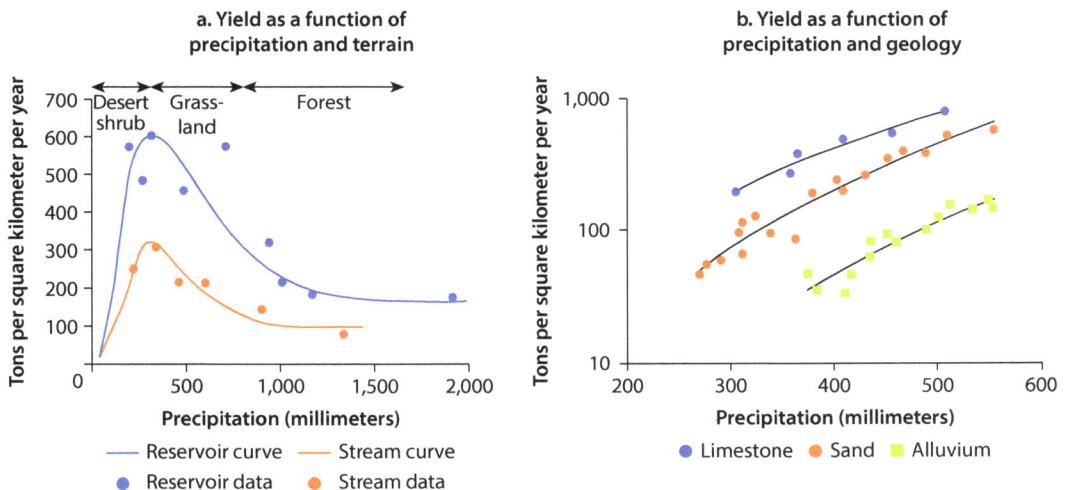

Sources: Adapted from Langbein and Schumm 1958 and Sharma and Chatterji 1982.

Figure 4.2 Approximate Ranges of Specific Sediment Yield for Various Regions in the United States

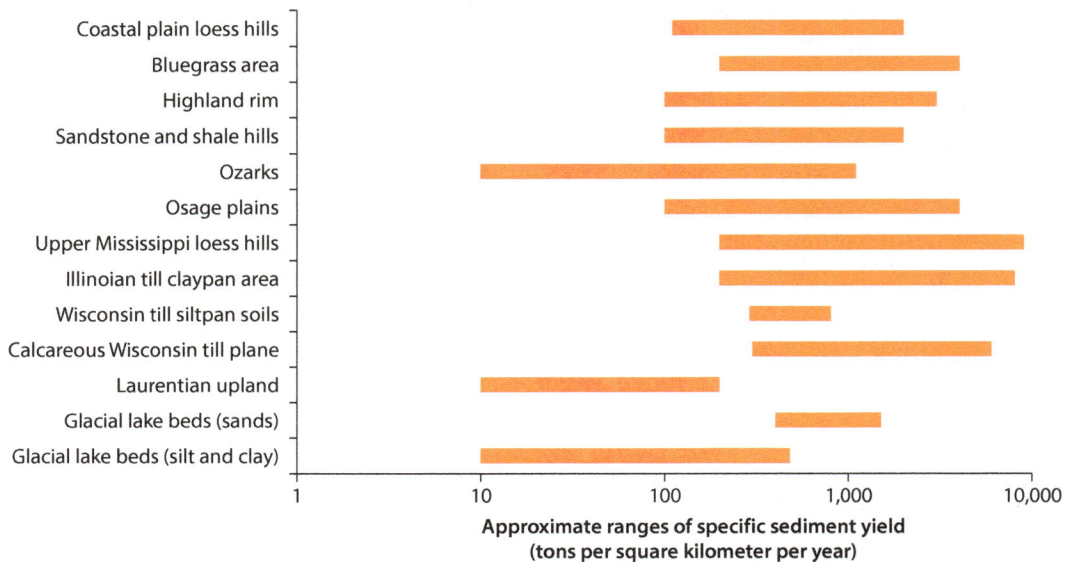

Approximate ranges of specific sediment yield
(tons per square kilometer per year)

Source: Adapted from Vanoni 1975.

would be to make use of field data, either concurrent measurements of sediment concentration and water flow data over long periods, or reservoir resurvey data providing estimates of the amount of sediment depositing in reservoirs over extended periods (see "Sediment Yield Estimation" section in this chapter). However, such data are often lacking during the planning and design phases of projects, requiring that empirical methods and sediment yield maps be used to accomplish this goal.

Estimates of sediment yield often differ significantly from what happens in real life. It is not uncommon to find that the actual sedimentation rate of a reservoir, that is, the rate at which a reservoir actually fills with sediment after construction, is much greater than estimates prepared before its construction. This experience further underscores the importance of dealing with this uncertainty when planning and designing dams and reservoirs. The most desirable method would be to provide and implement reservoir sediment management capabilities at dam and reservoir projects (see chapter 6). Such facilities should aim to protect the infrastructure against the uncertainties associated with sediment yield estimation, with a particular focus on making facilities available that will preserve reservoir storage space over the long term.

Human disturbance, such as farming, construction, and logging, increases erosion, as do forest fires. Forest fires are associated with high erosion resulting from loss of vegetation and the temporary presence of hydrophobic soils immediately after a fire. Hydrophobic soils are created when hydrocarbon residue forms after organic material is burned. It soaks into empty pore spaces in soils, making them impervious to water. This temporary imperviousness results in greater runoff when rain falls on the soils after forest fires, which in turn leads to increased erosion.

Megahan (1975) examines the impact of human disturbance on a 40-hectare mountain watershed in Idaho, United States, and finds that sediment yield can increase significantly because of such disturbance. Table 4.3 shows that disturbance in a watershed from tree felling, log skidding, and the construction of roads can result in orders of magnitude increases in sediment yield. Megahan (1975) finds that such activities increased sediment yield by about 155 times on average, compared with natural conditions. Mass erosion at roads resulted in even higher local increases in sediment yield, sometimes up to 550 times higher.

Erosion and Sediment Yield

Most of the sediment in rivers is generated in the headwaters because of a direct and close link between hill slopes and river channels in those locations. However, as rivers flow further downstream into valley floors and floodplains, the link between hill slopes and active river channels becomes more and more uncoupled. Sediment eroding from the channel banks and beds in a river becomes more dominant than the supply from hill slope colluvium. Although sediment eroded from hill slopes eventually reaches the outlet of a river to the ocean, it does so over a long period. The time needed for sediment generated in the headwater to reach the outlet depends on the distance to the outlet, the sediment size, and the frequency of storms.

Sediment is temporarily stored within the watershed on banks along rivers and in the alluvium in river channels. Sediment generated by erosion during a particular storm event may not enter a river until a later time, when another storm event may cause runoff transporting that sediment over land into the river channel. Generally, the hill slope erosion rate is larger than the sediment yield in a river. At any time, less sediment is carried by a river than what has eroded in the catchment.

If an estimate of the amount of erosion that may occur in a watershed is known, the sediment yield can be estimated by multiplying the erosion by a factor known as the sediment delivery ratio. An indication of how the sediment delivery ratio may change as a function of drainage area is shown in figure 4.3 for selected

Table 4.3 Increased Sediment Yield in a 40-Hectare Mountain Watershed

Type of disturbance	Sediment yield (cubic meters per square kilometer per year)	Ratio to undisturbed land
Undisturbed land	42	1
Average for disturbed watershed	6,325	155
Subwatershed by disturbance type		
Tree felling and log skidding only	65	2
Roads (surface erosion)	8,966	220
Roads (mass erosion)	22,417	550

Source: Megahan 1975.

Extending the Life of Reservoirs • http://dx.doi.org/10.1596/978-1-4648-0838-8

Figure 4.3 Sediment Delivery Ratio as a Function of Drainage Area

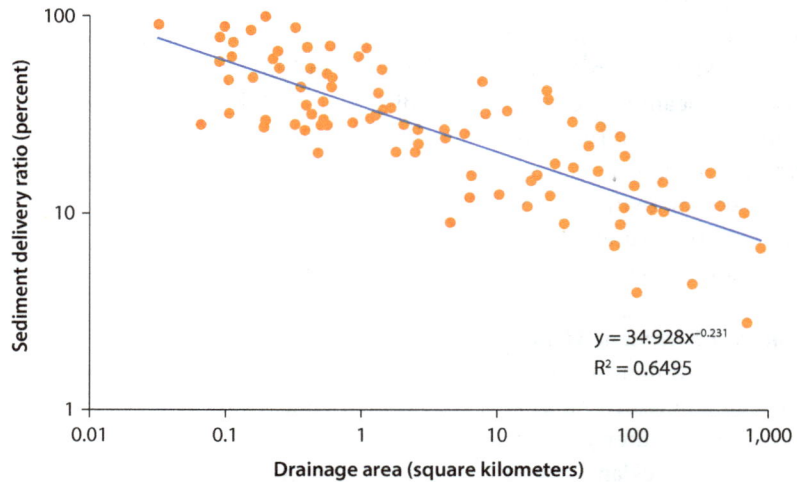

$$y = 34.928x^{-0.231}$$
$$R^2 = 0.6495$$

Source: Data from Boyce 1975.

regions in the United States. Note that the sediment delivery ratio decreases as the catchment area increases. This observation is in line with the description presented above. Temporary storage of eroded sediment in a catchment and river channel results in smaller amounts of sediment being transported by a river; that is, the sediment yield is generally lower than the erosion. For example, from figure 4.3 the sediment delivery ratio for a 200 square kilometer catchment is about 10 percent, meaning that the sediment yield from the area is equal to 10 percent of the total mass of erosion that may occur over the long term.

Temporal Variability

Temporal changes in sediment yield result from a number of factors, including the occurrence of extreme events (for example, storms, floods, and mass failure of hill slopes), changing climate, seasonable variability in flow, and long-term changes in watershed conditions. A single extreme event can generate a substantial portion of the overall sediment yield from a watershed in a very short period. Milliman and Meade (1983) measure a discharge of 50 million tons during a single flood in a river in 1969, which is more than 700 times the average annual sediment load of that river of 0.069 million tons per year.

A study by Meade and Parker (1985) for rivers in the United States finds that about 50 percent of the annual sediment load is discharged in only 1 percent of the time, and 90 percent occurs in only 10 percent of the time. Similar results have been found in other parts of the world, such as in Austria, Puerto Rico, and the United Kingdom. For example, figure 4.4 shows a cumulative distribution curve of suspended sediment yield in Río Tanamá, Puerto Rico, as a function of time. This figure shows that 1 percent of the days account

Figure 4.4 Ranked Cumulative Sediment Yield from Río Tanamá, Puerto Rico, as a Function of Time

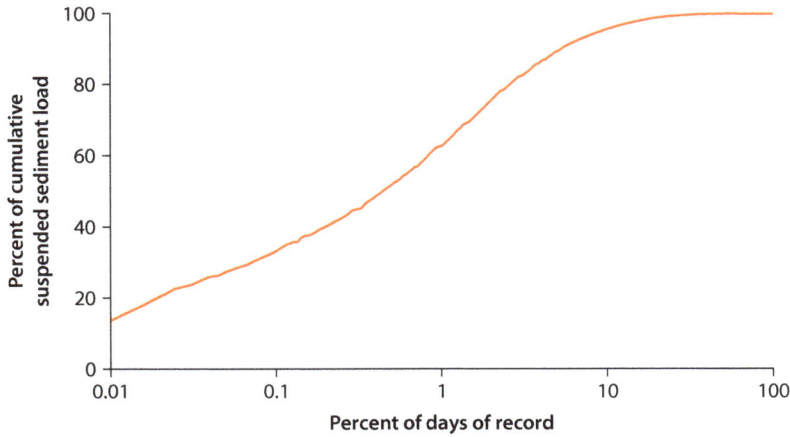

Source: Morris and Fan 1998.

for 67 percent of the sediment discharge over the 21-year sediment discharge record, and that 10 percent of the time accounts for about 95 percent of the total sediment load.

Long-term variability in the sediment load in a river may result from changes in land use (for example, agriculture to urban land use, or forest to crop), changes in climate, and reduced sediment supply caused by, for instance, soil conservation efforts in a catchment. Such long-term variability may be the result of a complex suite of processes, which are not always fully understood. Temporary storage of sediment generated under one set of conditions may not exist until later under a different set of conditions. Rooseboom (1992) finds that the rate of sediment discharge in the Orange River, South Africa, dramatically changed from about 1948 onward, and attributes the decline in sediment yield to upland sediment deposition in farm ponds and a decline in readily available supply of sediment due to historic soil loss (figure 4.5). As shown in figure 4.5, the slope of the curve is steeper before 1948 than after, indicating that sediment discharge was originally higher than in the subsequent decades.

Measuring Sediment Yield

The methods used to measure sediment yield consist primarily of river monitoring using concurrent measurements of water discharge and sediment concentrations in the flowing water, and repeated bathymetric surveys of reservoirs and lakes. The best estimates of average long-term sediment yield are obtained through the use of bathymetric surveys.

Extending the Life of Reservoirs • http://dx.doi.org/10.1596/978-1-4648-0838-8

Figure 4.5 Changes in the Rate of Sediment Discharge in the Orange River, South Africa, 1929–69

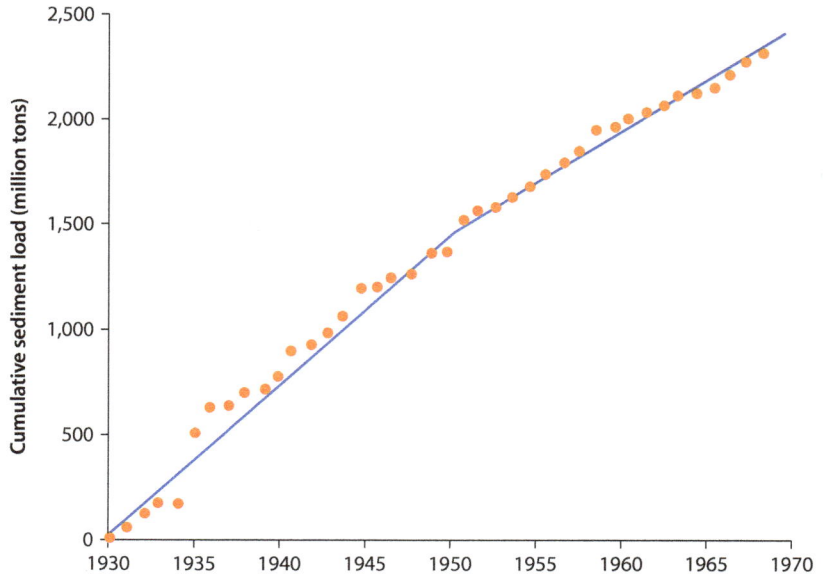

Source: Rooseboom 1992.

Bathymetric Surveys

Bathymetric surveys can be executed with relative ease thanks to advances in technology (see "Bathymetric Mapping of Sedimentation" in chapter 6). The most convenient way to execute bathymetric surveys is to use echosounders coupled with automatic global positioning systems. When a global positioning system receiver is mounted on a boat, a zig-zag path is followed across a water body to allow concurrent collection of sounding and positioning data, say every five seconds or so. The collected data can be saved in a simple database device for subsequent analysis to develop bathymetric maps of the reservoir or lake.

Comparison of the bathymetric map thus developed with the original bathymetry of the lake or reservoir allows the volume of sediment that has been deposited over a certain period to be quantified. Regular surveys at different times provide increased confidence in the volume of sediment deposited in a reservoir.

To convert the amount of sediment that has been deposited in the reservoir to the average sediment yield of the upstream watershed, the trap efficiency of the reservoir must be accounted for. Trap efficiency is the percentage of sediment flowing into a reservoir that is trapped and deposited in the reservoir. For example, if the trap efficiency of a reservoir is 90 percent, on average, 90 percent of the sediment flowing into that reservoir will be trapped and deposited, while about 10 percent will discharge downstream. Methods for estimating trap efficiency are presented in the "Trap Efficiency" section in chapter 5. Once the trap efficiency is known, the volume of sediment that originated from the watershed and flowed into the reservoir can be determined by dividing the estimated volume of deposited sediment by the trap efficiency.

As indicated, sediment yield is customarily presented in units of mass. Therefore, the bulk density of the deposited sediment in the reservoir under consideration must be estimated to convert the volume of sediment to mass. The bulk density can be estimated by either sampling the sediment deposited in a reservoir, which is difficult, or by making use of empirical techniques (see the section on "Sediment Bulk Density" in chapter 6). Once the bulk density of the deposited sediment has been estimated, the total mass of sediment that flowed into a reservoir can be calculated by multiplying the volume of sediment by the estimated bulk density.

Once the mass of sediment that entered the reservoir over a certain period (between bathymetric surveys) has been estimated, the specific sediment yield for the watershed can be quantified. The total estimated mass of sediment entering the reservoir is divided by the area of the catchment area and by the number of years between surveys. This calculation provides average specific sediment yield in $t/km^2/yr$.

River Monitoring

When monitoring rivers to estimate sediment load, water discharge and sediment concentration must be concurrently measured over relatively long periods. Table 4.4 demonstrates the importance of having long records available. Small sample sizes are particularly prone to error when the annual coefficient of variation[1] of the population from which the sediment or hydrology sample is taken is large. It is not unusual for hydrologic and sediment data in semi-arid and arid regions to have coefficients of variation on the order of 0.5, and even much higher.

A practical problem that often arises is that the high expense associated with collecting suspended sediment samples results in only a few years of data, if any, being available at many locations around the world. Records of only two

Table 4.4 Potential Error at 5 Percent Level of Significance for Various Sampling Sizes (Years) as a Function of Annual Coefficient of Variation

Potential error (%)	Population coefficient of variation					
	0.25	0.50	0.75	1.00	1.50	2.00
±100		1	2	4	9	15
±75		2	4	7	15	27
±50	1	4	9	15	35	62
±25	4	16	35	62	140	250
±10	24	96	216	385	865	1,538

Source: Annandale 1987.

Extending the Life of Reservoirs • http://dx.doi.org/10.1596/978-1-4648-0838-8

to four years are associated with potential errors in sediment yield estimation on the order of ±75 percent to ±100 percent. The importance of using as many different techniques as possible to estimate sediment yield is therefore clear, as indicated in the next section.

Sediment concentration measurements may be taken using standard equipment (Rasmussen et al. 2009; Interagency Water Resources Council 1940 [1967]) or more advanced equipment like the LISST sensors from Sequoia (www.sequoiasci.com), which use laser technology to concurrently measure sediment concentration and particle size distribution of suspended sediment.

Sediment Yield Estimation

General Approach

As noted, sediment yield depends on numerous factors, ranging from climate to geologic, topographic, and anthropogenic influences. Processes determining the sediment yield of a river are complex, making sediment yield estimation a difficult task best executed by experts.

Reliable sediment yield estimates use multiple methods and careful evaluation of the results. This is the most important characteristic of defensible sediment yield estimates. No one method used in isolation can be relied upon, not even the use of site-specific sediment rating curves. Relying upon only one method is undesirable because of the large variability inherent in sediment discharge.

For example, relying purely on sediment rating curves derived from sediment concentration and flow data may result in inaccurate estimates because the procedure may not, for example, fully account for the impact of anthropogenic influences, or it may not include sampling during high flow events when sediment load is at its highest.

The adverse outcomes of relying only on a sediment rating curve to estimate sediment yield has been amply demonstrated at Bakaru Hydropower Project, Sulawesi, Indonesia. Failure to consider other factors and how they may change in the future resulted in a one order of magnitude difference between the originally estimated and the actual sediment yield, leading to significant unplanned operational difficulties (personal communication, Bakaru staff).

Methods for estimating sediment yield, based both on field data and on empirical analysis, are discussed below.

Methods Based on Field Data

Sediment Rating Curves

River monitoring can provide data that can be used to develop sediment rating curves consisting of graphs relating suspended sediment concentration and water flow data (see "Sediment Rating Curves" section in chapter 6). Estimating sediment yield using sediment rating curves requires multiplying sediment concentrations obtained from a rating curve by water flow data over a set period, either historic or computer generated. The results of such an analysis may be expressed as specific sediment yield, that is, in $t/km^2/yr$. Specific sediment yield can be interpolated or extrapolated by multiplying it by the area of the catchment area upstream of the dam and reservoir under consideration.

Bed Load Estimation

Should better estimates of bed load be required, as may be the case for mountain rivers, they may be developed by making use of bed material properties (particle size distribution) and reliable theoretical equations. Approaches for collecting and determining bed material properties are presented in chapter 5. Methods developed by Parker (1990) and Wilcock and Crowe (2003) are often used to quantify bed load if bed material properties are known. When data are lacking, a rule-of-thumb approach often used is to increase the suspended sediment load estimate by a factor of 5 percent to 10 percent, and possibly up to 25 percent in some mountain rivers.

Bathymetric Surveys

If bathymetric survey data are available for the river of concern or similar surrounding rivers, these data can be used to estimate sediment yield, as explained in the "Bathymetric Surveys" section of this chapter. If historic bathymetric surveys are not available, new surveys may be commissioned on existing facilities and the results regionalized. If bathymetric surveys are available for a number of surrounding rivers, the specific sediment yield (expressed in $t/km^2/yr$) can be regionalized. The average annual sediment yield at a particular facility can then be determined by multiplying the specific sediment yield upstream of the dam and reservoir by the area of the catchment area.

Empirical Methods

Sediment monitoring and bathymetric data are often not concurrently available over long historic periods, making it impossible to base sediment yield estimates on field data. Under such circumstances, empirical techniques may be used to estimate average annual sediment yield. Of course, even if field data are available and can be used to estimate sediment yield as described above, the use of empirical techniques provide an additional perspective. Empirical methods that may be used are briefly summarized below.

Sediment Yield Maps

Sediment yield maps, such as those shown in map 4.1, may occasionally be available on a regional basis if previously developed through research projects. In such cases, sediment yield may be estimated by multiplying specific sediment yield (expressed in $t/km^2/yr$) by the area of the catchment area upstream of the dam and reservoir. In the absence of local sediment yield maps, map 4.1 may be used as guidance.

Extending the Life of Reservoirs • http://dx.doi.org/10.1596/978-1-4648-0838-8

Empirical Techniques

Practical experience by the authors of this book indicates that the empirical technique developed by Syvitski and Milliman (2007) usually provides defensible estimates of sediment yield in large catchments (several hundred square kilometers). The method accounts for geologic features, climate, anthropogenic influences, population density, level of development, and topography. Use of this method facilitates consideration of future catchment changes such as urban development, construction of dams, deforestation, and so on.

Geomorphologic Approaches

Sediment yield may be estimated by making use of geomorphologic approaches, which entail dividing a catchment into geomorphologically similar regions and proceeding from there to estimate sediment yield (see, for example, Kondolf, Rubin, and Minear 2014). Such approaches require the expertise of a specialized geomorphologist.

Computer Simulation

Computer simulation using Soil and Water Assessment Tool (SWAT) software may be used to estimate sediment yield. For such simulations to add value, the model must be calibrated using field data that have been collected with integrated samplers (see "Sampling for Suspended Sediment Load" in chapter 6). Conventional water quality samples, which are usually collected as grab samples with bottles close to the water surface, are inadequate. The use of water quality samples generally results in severe underestimates of sediment yield (Walling 2008).

Quantifying Sediment Yield

Once estimates of sediment yield have been prepared using as many techniques as possible, the sediment yield at a reservoir can be quantified by considering all the results. The values obtained from the different techniques normally differ, often substantially. Setting a representative sediment yield value using these results is no mean task, and it is usually necessary to rely on the judgment of experts in the field. Using a review consultant is in order.

Note

1. The annual coefficient of variation is equal to the annual standard deviation divided by the annual mean of the variable.

References

Annandale, G. W. 1987. *Reservoir Sedimentation*. New York: Elsevier Science Publishers.

Boyce, R. C. 1975. "Sediment Routing with Sediment-Delivery Ratios." In *Present and Prospective Technology for Predicting Sediment Yields and Sources*, ARS-S-40, 61–65. Oxford, MS: USDA Sedimentation Lab.

Interagency Water Resources Council. 1940 [1967]. "Field Practice and Equipment Used in Sampling Suspended Sediment." St. Paul Engineer District Sub-Office, Hydraulic Laboratory, University of Iowa, Iowa City, Iowa.

Kondolf, G. M., Z. K. Rubin, and J. T. Minear. 2014. "Dams on the Mekong: Cumulative Sediment Starvation." *Water Resources Research* 50 (6): 5158–69. doi:10.1002/2013WR014651.

Langbein, W. B., and S. A. Schumm. 1958. "Yield of Sediment in Relation to Mean Annual Precipitation." *Transactions, American Geophysical Union* 30 (6): 1076–84.

Martin, J-M., and M. Meybeck. 1979. "Elemental Mass-Balance of Material Carried by Major World Rivers," *Marine Chemistry* 7 (3): 173–206.

Meade, R. H., and R. S. Parker. 1985. "Sediment in Rivers of the United States." In *National Water Summary 1984*, U.S. Geological Survey Water-Supply Paper 2275. Washington, DC: U.S. Government Printing Office.

Megahan, W. F. 1975. "Sedimentation in Relation to Logging Activities in the Mountains of Central Idaho." In *Present and Prospective Technology for Predicting Sediment Yields and Sources*, ARS-S-40, 74–82. Oxford, MS: USDA Sedimentation Lab.

Milliman, J. D., and R. H. Meade. 1983. "World-Wide Delivery of Sediment to the Oceans." *Journal of Geology* 91 (1): 1–21.

Morris, G. L., and J. Fan. 1998. *Reservoir Sedimentation Handbook: Design and Management of Dams, Reservoirs and Watersheds for Sustainable Use*. New York: McGraw-Hill.

Parker, G. 1990. "Surface-Based Bedload Transport Relation for Gravel Rivers." *Journal of Hydraulic Research* 28 (4): 417–36.

Rasmussen, P., J. R. Gray, G. D. Glysson, and A. C. Ziegler. 2009. *Guidelines and Procedures for Computing Time-Series Suspended-Sediment Concentrations and Loads from In-Stream Turbidity-Sensor and Streamflow Data*. Reston, VA: United States Geological Survey.

Rooseboom, A. 1992. "Sediment Transport in Rivers and Reservoirs: A South African Perspective." Report to Water Research Commission of South Africa, by Sigma Beta Consulting Engineers, Stellenbosch.

Sharma, K. D., and P. C. Chatterji. 1982. "Sedimentation in Nadis in the Indian Arid Zone." *Hydrological Sciences Journal* 27: 345–52.

Syvitski, J. P. M., and J. D. Milliman. 2007. "Geology, Geography, and Human Battle for Dominance over the Delivery of Fluvial Sediment to the Coastal Ocean." *Journal of Geology* 115: 1–19.

Vanoni, V. A. 1975. *Sedimentation Engineering*. Reston, VA: American Society of Civil Engineers.

Walling, D. E. 1987. "Rainfall, Runoff and Erosion of the Land: A Global View." In *Energetics of Physical Environments*, edited by K. J. Gregory, 89–117. New York: John Wiley and Sons.

———. 2008. "The Changing Sediment Load of the Mekong River." *Ambio* 37 (3): 150–57.

Walling, D. E., and B. W. Webb. 1983. "Patterns of Sediment Yield." In *Changing River Channels*, edited by K. I. Gregory. Chichester, U.K.: Wiley.

Wilcock, P. R., and J. C. Crowe. 2003. "Surface-Based Transport Model for Mixed-Size Sediment." *Journal of Hydraulic Engineering* 129 (2): 120–28.

CHAPTER 5

Patterns of Sediment Transport and Deposition

George W. Annandale

Introduction

When studying reservoir sedimentation, it is necessary to estimate the amount of sediment that will be deposited in a reservoir, the shape of the deposit, and how it will change with time. This chapter demonstrates that two well-known empirical techniques for estimating the amount of sediment that will be deposited in a reservoir produce estimates comparable to more advanced computer simulation modeling. The shapes of deposited sediment in reservoirs and how the sediment sizes of deposited sediment may change are also discussed, followed by an example of how a reservoir may reach a new equilibrium during advanced stages of reservoir sedimentation.

Sediment Transport in Reservoirs

Sediment can be transported in reservoirs by two modes: conventional sediment transport and density currents. Density currents occur when the density of the sediment-water mixture discharging into the reservoir is much higher than the density of the clean water contained in the reservoir. When this happens, the high-density water containing the sediment may travel along the reservoir bed in the form of a distinct current (sometimes referred to as a turbidity current). Such a current may travel long distances, eventually reaching the dam. Density currents traveling approximately 100 kilometers along the bed of a reservoir are observed on an annual basis at Xiaolangdi Dam, China. The density currents usually contain fine to very fine sediment.

Conventional sediment transport occurs when the turbulence in water flowing through a reservoir carries suspended sediment and bed load (coarse sediment particles located on a river bed) into a reservoir. The suspended sediment is distributed throughout the entire water column (not only along the bed as with density currents). When bed load is transported into a reservoir, closely hugging the reservoir bed, it is generally deposited in totality upstream of the dam. Of the suspended load consisting of sand, silt, and clay, some may deposit while a portion may travel through the reservoir for release downstream of the dam.

Trap Efficiency

As a river carrying sediment flows into a reservoir, the water slows down causing a decrease in sediment transport capacity, resulting in sediment depositing in the reservoir. Some of this sediment may pass through the reservoir for release downstream, meaning that only a portion of the sediment may be deposited in the reservoir. A variable known as trap efficiency (TE) expresses how much of the inflowing sediment may be trapped in a reservoir. TE is defined as the amount of sediment depositing in a reservoir divided by the total amount of sediment discharging into it.

Two empirical methods that can be used to estimate the TE of a reservoir are the Brune curve (1953) and the Churchill curve (1948). The Brune curve (figure 5.1) relates TE to the average annual residence time in a reservoir, while the Churchill curve (figure 5.2) relates the amount of sediment passing through a reservoir to a sedimentation index. The Churchill curve shown in figure 5.2 has been modified by Roberts (1982) by converting the sedimentation index to a dimensionless parameter (Annandale 1987).

The relative size of a reservoir is determined by dividing the storage volume of the reservoir by the mean annual flow volume that discharges into the reservoir. This ratio is known as the capacity-inflow ratio, as shown on the horizontal axis in figure 5.1. The Brune curve shows that relatively small

Figure 5.1 Brune Curve for Estimating the Trap Efficiency of Reservoirs

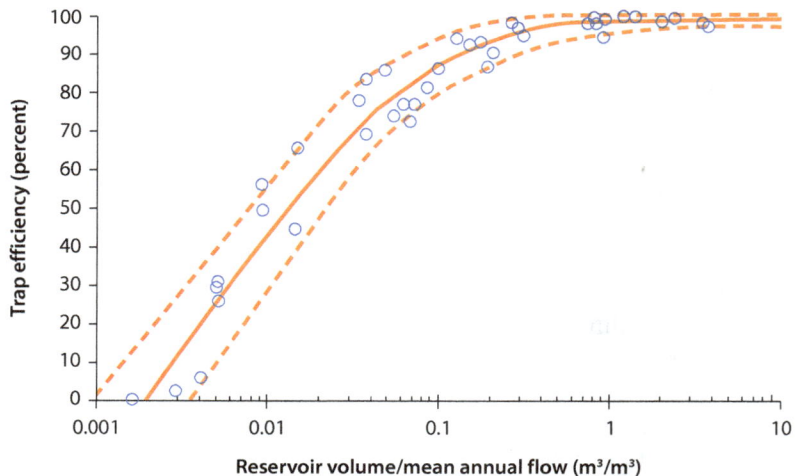

Source: Brune 1953.

Figure 5.2 Churchill Curve Modified by Roberts (1982) as Reported in Annandale (1987)

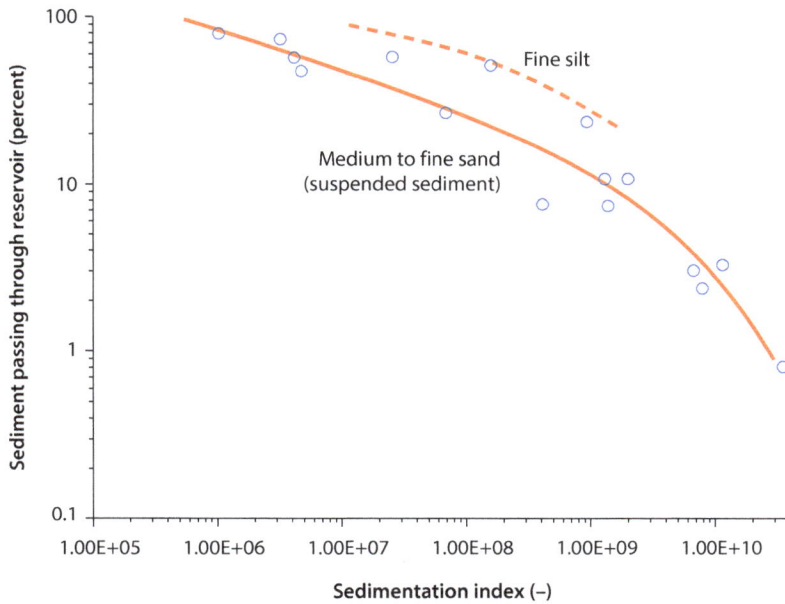

Source: Annandale 1987.

Note: Sedimentation index = $\dfrac{\text{(acceleration of gravity) (reservoir volume)}^2}{\text{(discharge)}^2 \text{ (reservoir length)}}$

reservoirs capture significant amounts of sediment. For example, a reservoir with a relative volume of only 0.1 captures approximately 85 percent of the sediment flowing into it.

The sedimentation index along the horizontal axis of the Churchill curve (figure 5.2) provides a rough indication of the residence time and the average flow velocity through a reservoir, relating it to the percentage of sediment passing through a reservoir, as plotted along the vertical axis of the figure.

The Brune and Churchill curves provide relatively consistent estimates of the TE of reservoirs. Figure 5.3 compares estimates of TE for a reservoir in Costa Rica using the Brune and Churchill curves and a computer simulation with the MIKE 21C software. The figure shows that the three curves are virtually identical, indicating good correlation between the three approaches. From a practical point of view, it can be concluded that the two empirical methods provide good estimates, requiring much less effort than computer simulation.

Spatial Distribution of Deposited Sediment

The classic shape of deposited sediment in a reservoir is shown in figure 5.4. The entire deposition pattern consists of topset, foreset, and bottomset beds. The topset and foreset beds consist mainly of coarse sediment and are known as

Figure 5.3 Alternative Sediment Trap Efficiency Estimates for a Reservoir in Costa Rica

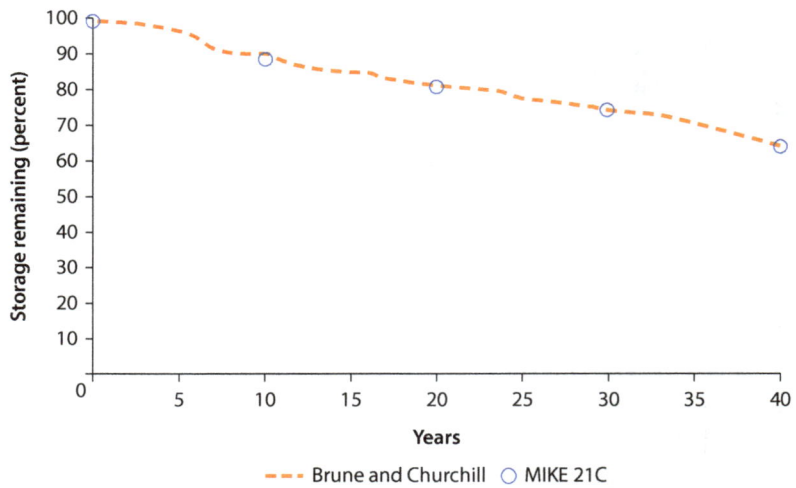

Figure 5.4 Typical Shape of Deposited Sediment in a Reservoir

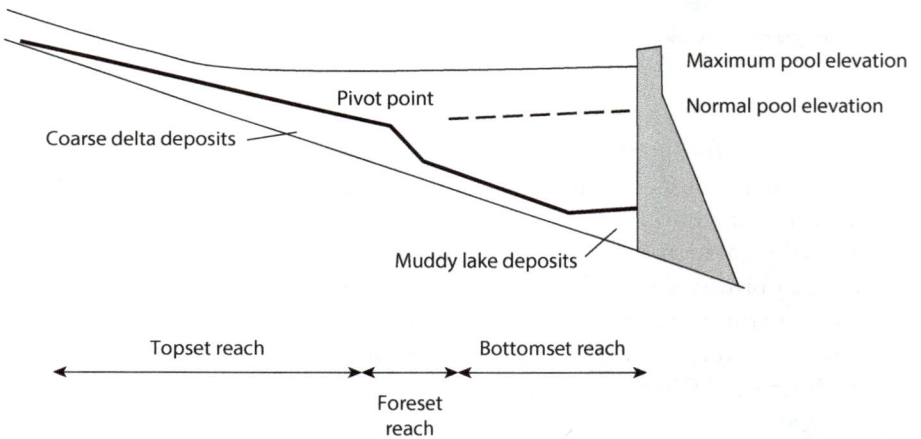

the delta, with its characteristic pivot point. The bottomset portion of the deposit generally consists of fine sediment, with a muddy lake sometimes forming just upstream of the dam. A muddy lake is an indication that density currents may be present in the reservoir.

Figure 5.4 shows two reservoir water surface elevations: maximum pool elevation and normal pool elevation. The maximum pool elevation is the water surface elevation that is not exceeded during operation of the reservoir. The normal pool elevation is the water surface elevation that the dam operator attempts to maintain most of the time. It is commonly believed that the location of the pivot point is influenced by the normal pool elevation in the reservoir, as shown in figure 5.4.

Note also that the upstream end of the deposited sediment is located upstream of the maximum pool elevation. This deposition results from the backwater that forms when floods enter a full reservoir. The backwater slows down the incoming flow in the portion of the river reach that is higher than the maximum water surface elevation, thereby depositing coarse sediment in that reach of the river.

Four variations of sediment deposit shapes are shown in figure 5.5, categorized as delta, tapering, wedge, and uniform distributions. These representations are somewhat simplified given that they often occur concurrently. For example, the wedge-like deposit that results from density currents often occurs in conjunction with the other three shapes illustrated in figure 5.5.

However, it is important to note that the wedge shape rarely occurs on its own. When establishing dead storage space, designers often incorrectly assume a wedge shape deposition pattern. The occurrence of such a deposition pattern is actually extremely rare, and principally results from a predominance of density currents and the absence of any coarse sediment entering a reservoir.

The other three shapes (delta, tapering, and uniform) are largely determined by the particle size distributions of incoming sediment, flood occurrence, whether density currents exist, and how the reservoir is operated. The shape of the deposited sediment may be best determined during the design process through computer simulation, although empirical techniques exist to accomplish this task.

The general shape and progression of deposited sediment in a reservoir (figure 5.4) has profound consequences for dam design. The storage space upstream of a dam is normally divided into active and dead storage space (figure 5.6).

Figure 5.5 Differing Shapes of Deposited Sediment in Reservoirs

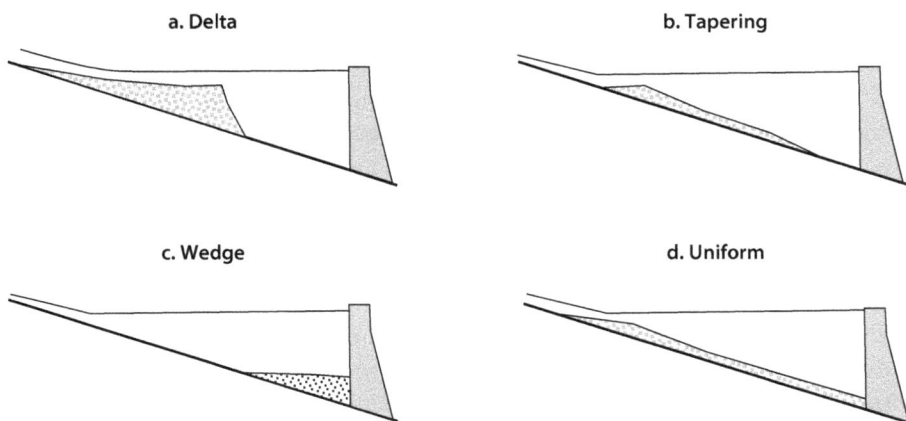

Source: Based on Morris and Fan 1998.

Figure 5.6 Active and Dead Reservoir Storage

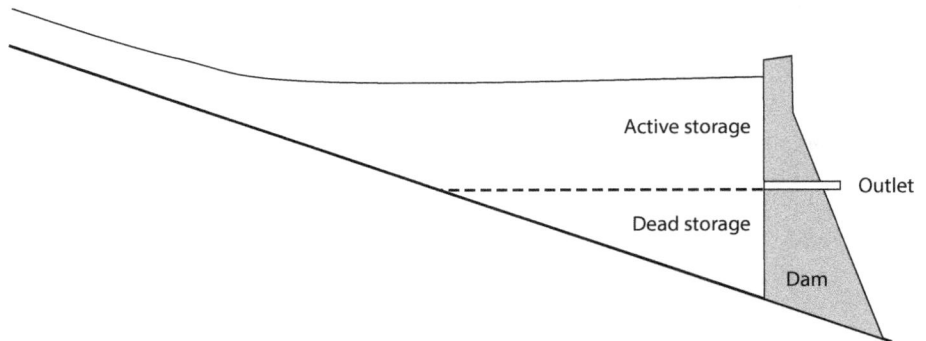

The dead storage space is located below the elevation of the invert of the low-est outlet in a dam. It is referred to as dead storage space because it is not possible to use it unless a special effort is made to pump the water from it. Otherwise, water cannot be released from the dead storage space because no outlets are low enough to do so. The active storage space shown in figure 5.6 stores water for supply, power generation, or flood control. The volume of water in the active storage space can be controlled by the outlet(s) in the dam.

In the case of run-of-river schemes, the active storage space may be very small compared with the dead storage space. However, the vast majority of large dams in the world are storage facilities where the active storage spaces are significantly larger than the dead storage spaces. Of all large dams, 73 percent were built to provide storage (37 percent for irrigation, 7 percent for hydropower, 15 percent for domestic water supply, and 14 percent for flood control) and 11 percent are run-of-river hydropower facilities (ICOLD 2015).[1]

Uninformed views of the spatial distribution of deposited sediment in a reservoir, which still exist to this day, assume that sediment is deposited close to the dam, in horizontal layers, gradually filling the reservoir from the bottom up. Based on this assumption, dam designers erroneously allocate "dead storage space" in the belief that all sediment entering the reservoir will deposit only in this space and that the active storage space will be available for use over an extended period. Based on this incorrect assumption, a designer may calculate the total volume of sediment that will deposit in a reservoir over a certain period and incorrectly equate the dead storage volume to the total estimated volume of deposited sediment.

This incorrect view of reservoir sedimentation has significant consequences for dam design and for the long-term performance of a reservoir. This issue is discussed in "Storage Loss and its Impacts" in chapter 3, where figure 3.2 illustrates that deposition of sediment in the active storage zone (due to delta formation) plays an important role in diminishing power production efficiency (figure 3.3).

Modern dam designs must account for the impact of the actual shape of deposited sediment on reductions in the *active* reservoir storage space. Failure to recognize these patterns (figures 3.2 and 5.4) leads to economically inferior designs and incorrect impressions of the long-term performance of the reservoir storage space for water supply, power generation, and flood control.

Empirical Techniques

The shape of the deposited sediment may best be determined by using computer simulation, although empirical techniques do exist.

Empirical determination of the shape of deposited sediment may be used during the prefeasibility stages of a project. However, during feasibility studies computer simulations must be used to establish the dimensions of the dam and reservoir.

Estimating Topset Slope

Two empirical methods that can be used to quantify the topset slope of a delta are shown in figures 5.7 and 5.8. Strand and Pemberton (1987) plot the topset slope of existing reservoirs against the original riverbed slope and find that the topset slope can range from equal to the original riverbed slope to about 20 percent of the original riverbed slope (figure 5.7). Menne and Kriel (1959) follow a slightly different approach. They relate the ratio of the topset to the original riverbed slope to a shape factor (defined as the ratio of the reservoir length to the average reservoir width). Their estimates of the ratio between the topset and original riverbed slopes (indicated by the solid dots in figure 5.8) range between 70 percent and 20 percent of the original riverbed slope.

Menne and Kriel (1959) provide information as to how the topset slopes are related to the general shape of a reservoir. Figure 5.8 indicates that long, thin reservoirs (high ratio between the reservoir length and average reservoir width) have milder topset slope, while the slope becomes steeper as the reservoir becomes relatively wider, which, in principle, makes sense.

The blue dots in figure 5.8 were subsequently added by Annandale (1987) for South African reservoirs, indicating that topset slopes may often equal the original riverbed slope, essentially representing either the uniform or the tapering sediment distributions identified in figure 5.5. The deposited sediment in the reservoirs investigated by Annandale (1987) principally consists of medium to fine sand and silt, potentially indicating the presence of density currents that are depositing sediment in a more gradual way as it travels through a reservoir.

The foreset slope of a delta is deemed to equal 6.5 times the topset slope, as recommended by Strand and Pemberton (1987). Qian (1982) finds that the foreset slope equals 1.6 times the original riverbed slope.

Estimating Spatial Distribution of Deposited Sediment
Borland and Miller (1960)
Based on analysis of reservoir sedimentation data in the United States, Borland and Miller (1960) identify four types of reservoirs. The data indicate

Extending the Life of Reservoirs • http://dx.doi.org/10.1596/978-1-4648-0838-8

Figure 5.7 Relationship between the Topset Slope of a Delta and the Original Riverbed Slope for Existing Reservoirs

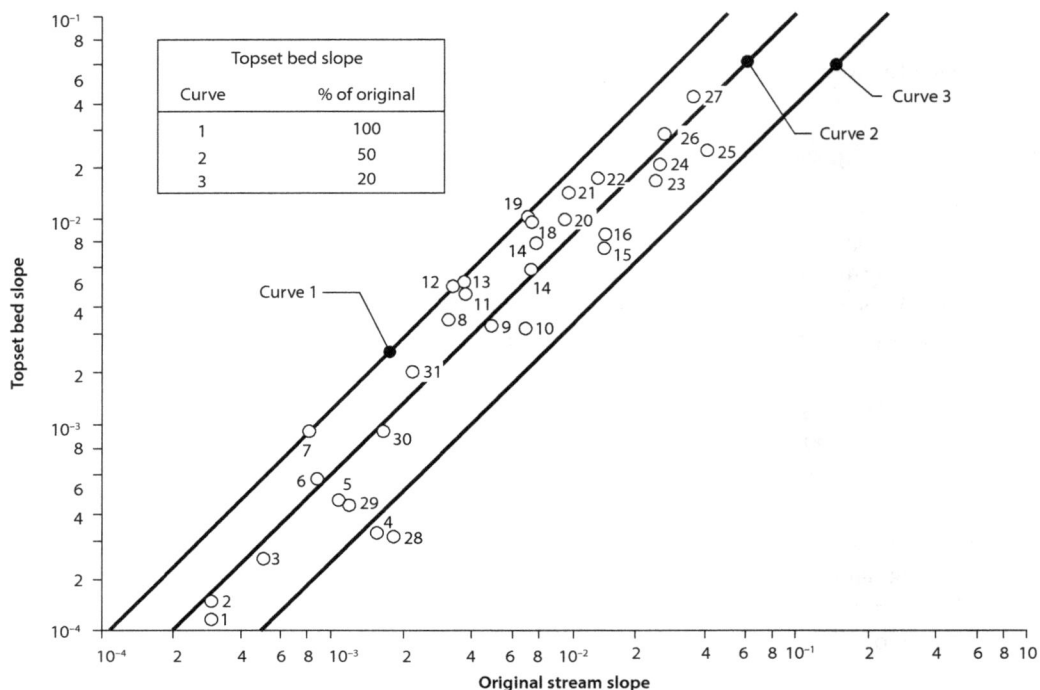

LOCATION OF PLOTTED POINTS

Point No.	Reservoir	Stream
1	Dennison	Washita and Red River
2	Dennison	Washita and Red River
3	McMillan	Pecos River
4	Avalon	Pecos River
5	Guernsey	North Platte River
6	Altus (67 survey)	North Fork Red River
7	Altus (48 survey)	North Fork Red River
8	Boysen	Fivemile Creek (Wyoming)
9	Verdi Diversion Dam	Truckee River
10	Soeltzer	Clear Creek (California)
11	Jemez Canyon (65 survey)	Jemez River
12	Boysen	Muddy Creek (Wyoming)
13	Jemez Canyon (59 survey)	Jemez River
14	Jones Creek	Missouri River Loess Hills
15	Keswick (59 survey)	Spring Creek (California)
16	Keswick (57 survey)	Spring Creek (California)
17	Tongue River	Tongue River
18	Elephant Butte	Rio Grande
19	Soeltzer	Clear Creek (California)
20	Alfred Lage	Missouri River Loess Hills
21	C. A. Stiles	Missouri River Loess Hills
22	Sheep Creek	Sheep Creek
23	Thomos Hodkin	Missouri River Loess Hills
24	Fred Brown	Missouri River Loess Hills
25	William Esbeck	Missouri River Loess Hills
26	Fred Hoorah	Missouri River Loess Hills
27	Emma La Frontz	Missouri River Loess Hills
28	Lake Mead	Colorado River
29	Angostura	Cheyenne River
30	Conchas	Canadian River
31	Theodore Roosevelt	Salt River

Source: Strand and Pemberton 1987.

Figure 5.8 Relationship between Topset Sediment Slope and the Shape Factor

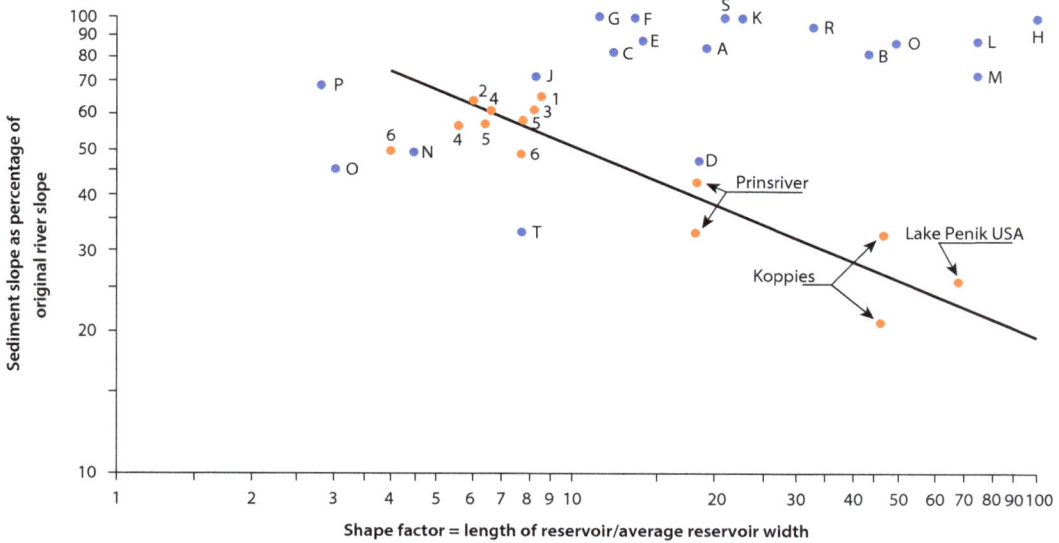

Source: Menne and Kriel 1959.
Note: Relationship between topset sediment slope and the shape factor = length of reservoir divided by average reservoir width.

a relationship between the reservoir shape and the percentage of sediment deposited at various depths throughout the reservoir. The type of reservoir can be related to the reciprocal value (M) of the slope of the line obtained by plotting reservoir depth on the vertical axis and reservoir capacity on the horizontal axis on a log-log scale.

The four standard types are (figure 5.9):

- Lake type I; M = 3.5–4.5; greater portion of the sediment is deposited in the upper part of the reservoir.
- Floodplain-foothill type II; M = 2.5–3.5.
- Hill type III; M = 1.5–2.5.
- Gorge type IV; M =1.0–1.5; greater portion of the sediment is deposited in the deeper part (dead storage zone) of the reservoir.

An example of how this method can be used to quantify the spatial distribution of deposited sediment may be found in Annandale (1987) and in Morris and Fan (1998).

Annandale (1987)
Using the data in figure 5.8 (blue dots), Annandale (1987) finds that the distribution of deposited sediment for those reservoirs, that is, tapering or uniform distributions, is related to the rate of change in the width of the reservoir from

Figure 5.9 Sediment Distribution in Four Reservoir Types

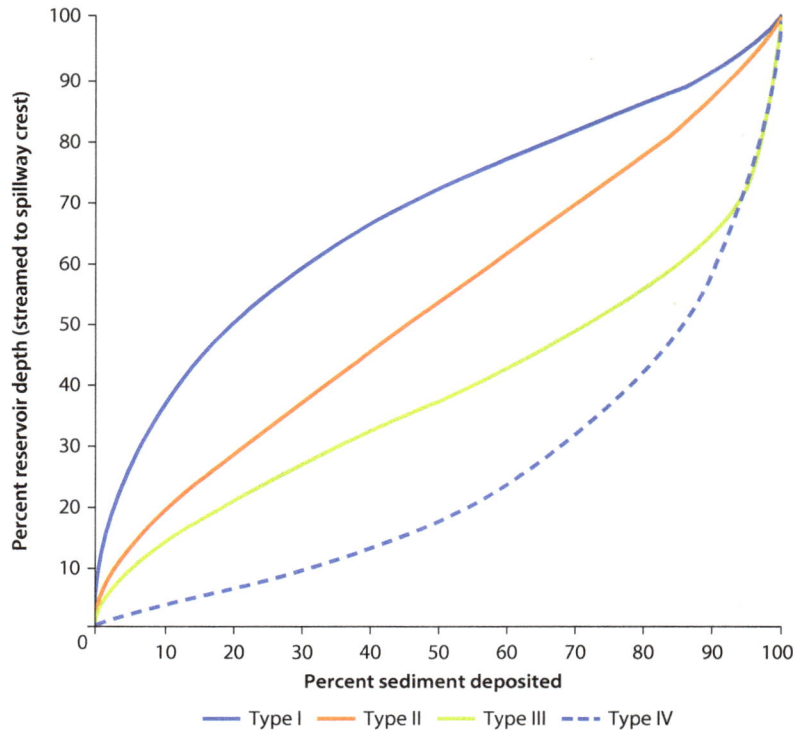

Source: Borland and Miller 1960.
Note: Type I = lake type; Type II = floodplain-foothill type; Type III = hill type; Type IV = gorge type.
See main text.

upstream to downstream. This rate of change is expressed by the following
equation:

$$\frac{dP}{dx} \approx \frac{P}{x}. \tag{5.1}$$

This represents the average ratio between the width of the reservoir P and
distance x measured from the upstream end of the reservoir. Equation (5.1) is
estimated by plotting width as a function of distance on a graph and obtaining
the average slope P/x.

Figure 5.10 shows how the total sediment volume in a reservoir is distributed
as a function of the dimensionless distance along the reservoir, measured from
the dam for various values of P/x. The figure indicates that sediment is less uniformly distributed if the reservoir is relatively wide, and that it is more uniformly
distributed when the reservoir is narrow. In principle this makes sense: the sediment transport capacity in a narrow reservoir will remain relatively high
throughout, thereby transporting sediment further into the reservoir, and vice

Figure 5.10 Dimensionless Cumulative Mass Curve Explaining Distribution of Deposited Sediment in a Reservoir

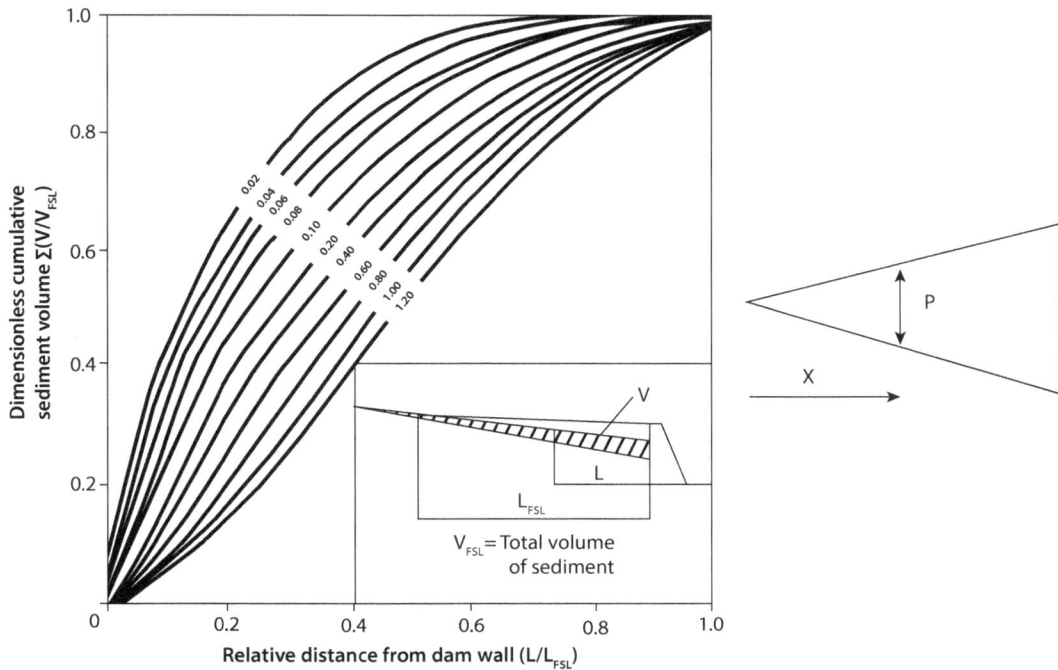

Source: Annandale 1987.
Note: FSL = full supply level; L = variable length measured from dam; P = width of reservoir at variable distance x from upstream end of reservoir; V = volume of sediment between the dam and location L; L_{FSL} = total length of the reservoir at full supply level; V_{FSL} = total volume of sediment underneath the full supply level.

versa for a relatively wide reservoir. For a wide reservoir, relatively more sediment will be deposited in the upstream reaches.

Figure 5.11, using the same nomenclature as figure 5.10, provides an indication of how much sediment might be deposited upstream of the maximum water surface elevation in a reservoir. For the reservoirs considered, the amount of sediment deposited upstream of the maximum water surface elevation equals about 4 percent of the total amount of sediment deposited in the reservoir and it is mainly deposited over a distance equaling about 20 percent of the reservoir length in an upstream direction.

Van Rijn (2013)

Van Rijn (2013) develops a method for empirically estimating the distribution of deposited sediment in the longitudinal direction of reservoirs. This is accomplished by quantifying the TE in predefined reaches of a reservoir. The TE is formulated as in equation (5.2):

$$E_{res} = 1 - \exp[-L \times A_{vr} \times (h - h_0) / h_2], \tag{5.2}$$

in which L = length of reservoir, h_0 = flow depth at upstream reservoir boundary (x = 0 meters), h = mean flow depth of reservoir (or section of reservoir) (figure 5.12), $A_{vr} = \alpha s(ws/u^*)(1+2 \ ws/u^*)$ = deposition parameter, $\alpha s = 0.25$ = coefficient (in range of 0.2–0.3), ws = settling velocity of sediment, u^* = mean bed shear velocity in reservoir. Comparison of calculation results using this technique was found to reasonably represent actual reservoir surveys (van Rijn 2013).

Figure 5.11 Distribution of Deposited Sediment above Full Supply Level in a Reservoir

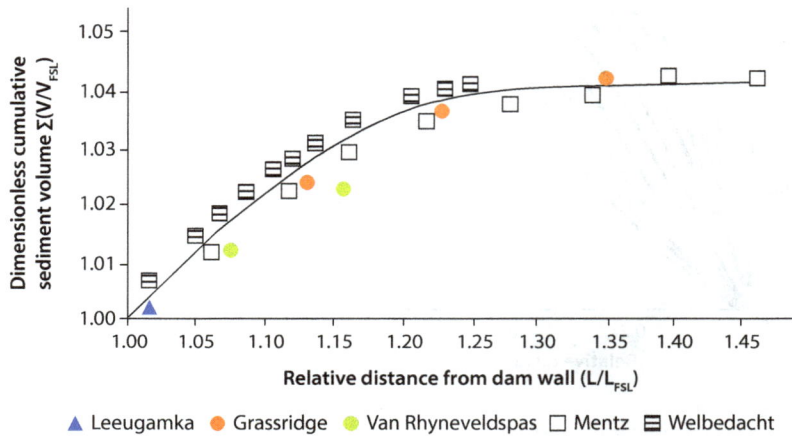

Source: Annandale 1987.
Note: L = variable length measured from dam; V = volume of sediment between the dam and location L; L_{FSL} = total length of the reservoir at full supply level; V_{FSL} = total volume of sediment underneath the full supply level.

Figure 5.12 Schematization of Reservoir Compartments Used to Estimate Distribution of Deposited Sediment

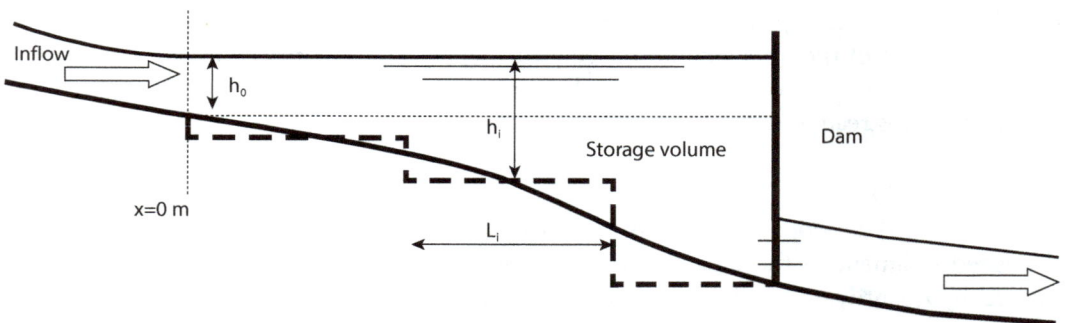

Source: van Rijn 2013.
Note: h_i = average water depth in section i; h_0 = water depth at upstream end of reservoir; L_i = length of section i; x = variable distance from upstream end of reservoir.

Computer Simulation

Although the empirical techniques outlined in the previous section may be used in prefeasibility studies, computer simulations of sediment deposition are preferable during feasibility studies and final design. Experience has shown that unsteady, non-uniform flow modeling approaches should be followed during computer simulations of reservoir sedimentation, particularly if a long time series with fluctuating water levels (due to opening and closing of outlets at the dam) are simulated. It has been found that stepwise, steady, non-uniform flow modeling (sometimes referred to as quasi-unsteady state modeling) creates mass balance problems during the simulation, which lead to incorrect simulation results. Computer simulation of reservoir sedimentation is specialized, and more discussion is outside the scope of this book.

Computer software that may be used to simulate sediment deposition includes the latest versions of HEC-RAS (using unsteady flow simulation), MIKE 11, MIKE 21C, and others. HEC-RAS and MIKE 11 are one-dimensional models that provide reasonable results when using unsteady flow simulation in narrow reservoirs. MIKE 21C is a quasi-three-dimensional model that provides good results in all cases, but is particularly useful for wide reservoirs.

Particle Size Distributions of Deposited Sediment

When a river flows into a reservoir, it slows down and deposits coarse sediment first, followed by finer sediments as it flows further downstream into the reservoir. Figure 5.13 provides an example of how the particle size distribution of deposited sediment changes. Sediment discharges into this reservoir in the form of bed load, suspended sediment load, and as density currents. The sediment

Figure 5.13 Particle Size Distributions from Four Locations in Sakuma Reservoir after 24 Years of Operation

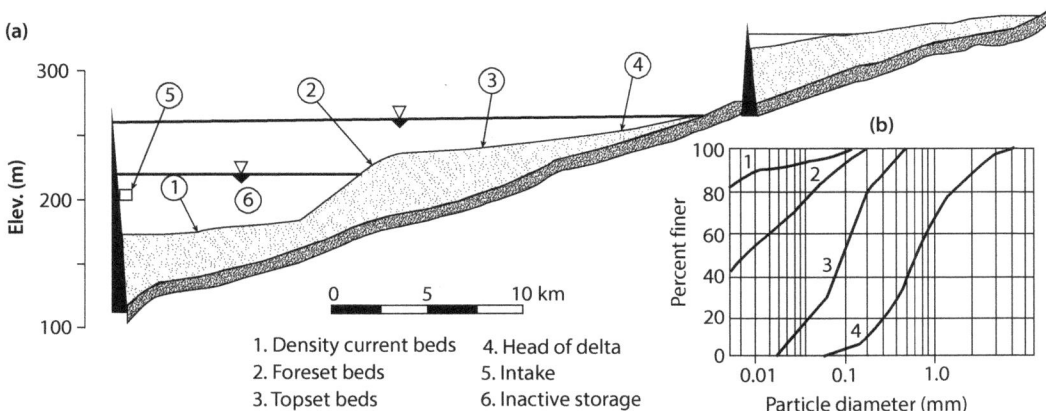

1. Density current beds
2. Foreset beds
3. Topset beds
4. Head of delta
5. Intake
6. Inactive storage

Sources: Okada and Baba 1982; Okumura and Sumi 2013.
Note: The longitudinal section of the reservoir shows both the delta and turbidity current deposits.

Extending the Life of Reservoirs • http://dx.doi.org/10.1596/978-1-4648-0838-8

contained in density currents is principally deposited close to the dam in a muddy lake, while the remainder of the sediment deposits in the topset and foreset slopes of the delta. The figure identifies the locations of the muddy lake (density current deposits) and the foreset and topset beds. To the right of the figure are shown particle size distributions of deposited sediment at four locations. Note that the sediment sizes for the density current deposits are very fine, and that particle size increases with increasing distance upstream.

Temporal Aspects of Sediment Deposition

As sediment deposits in a reservoir, the storage space eventually reaches a new equilibrium. The relationship between reservoir storage and time as deposition of sediment continues can be predicted by iteratively using the Brune and Churchill curves, as alluded to in figure 3.12. Reservoir resurveys of actual reservoir sedimentation, such as for Welbedacht Dam in South Africa, confirm the general progression of reservoir storage volume to a new equilibrium. The Welbedacht Dam reservoir has silted up substantially since it was originally commissioned in 1973. Figure 5.14 shows the progressive reduction in the reservoir's volume, stabilizing roughly from 1995 onward. This stabilization implies that the

Figure 5.14 Change in Reservoir Storage Volume Due to Reservoir Sedimentation for Welbedacht Dam, South Africa, from Commissioning to 2003

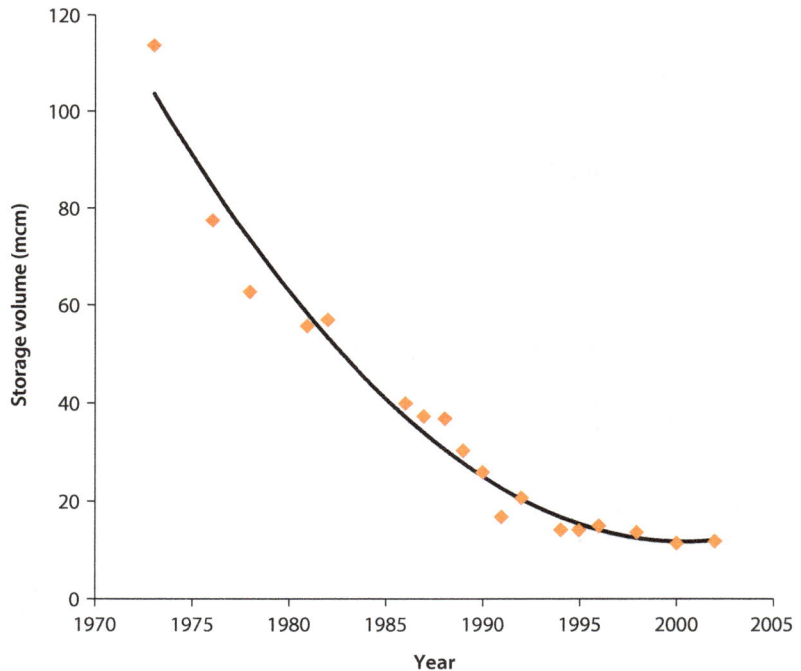

Source: Based on data from de Villiers and Basson 2007.

reservoir has reached a new geomorphic equilibrium, and that the amount of sediment flowing into it on average is roughly equal to the amount of sediment flowing out of it on an annual basis. For Welbedacht Dam, stable conditions were reached in roughly 20 years. The trend toward equilibrium has been influenced by sluicing operations every year since 1990. Sluicing in the case of Welbedacht Dam entails drawing down the reservoir water surface elevation at the dam during large flood events for just a few hours. The objective is to pass as much sediment through a reservoir as possible, without deposition.

Note

1. The remaining dams are used for recreation, navigation, fisheries, and other uses.

References

Annandale, G. W. 1987. *Reservoir Sedimentation*. New York: Elsevier Science Publishers.

Borland, W. M., and C. R. Miller. 1960. "Distribution of Sediment in Large Reservoirs." *Transactions of the American Society of Civil Engineers* 125: 160–80.

Brune, G. M. 1953. "Trap Efficiency of Reservoirs." *Transactions of the American Geophysical Union* 34 (3): 407–18.

Churchill, M. A. 1948. "Analysis and Use of Reservoir Sedimentation Data." Discussion of paper by L.C. Gottschalk in *Federal Interagency Sedimentation Conference, Proceedings*, 139– 40. Denver, CO: U.S. Geological Survey.

de Villiers, J. W. L., and G. R. Basson. 2007. "Modeling of Long-Term Sedimentation at Welbedacht Reservoir, South Africa." *Journal of the South African Institution of Civil Engineering* 49 (4): 10–18.

ICOLD (International Commission on Large Dams). 2015. "Register of Dams." http://www .icold-cigb.org/GB/World_register/general_synthesis.asp.

Menne, T. C., and J. P. Kriel. 1959. "Determination of Sediment Loads in Rivers and Deposition of Sediment in Storage Reservoirs." Technical Report 3, Department of Water Affairs, South Africa.

Morris, G. L., and J. Fan. 1998. *Reservoir Sedimentation Handbook*. New York: McGraw-Hill.

Okada, T. and Baba, K. 1982. Sediment release plan at Sakuma Reservoir, Proc. 14th ICOLD Congress, Rio de Janeiro.

Okumura, H., and T. Sumi. 2013. "Reservoir Sedimentation Management in Hydropower Plant Regarding Flood Risk and Loss of Power Generation." Proceedings of the International Symposium on Dams for a Changing World, The 80th Annual Meeting of ICOLD.

Qian, N. 1982. Reservoir sedimentation and slope stability: technical and environmental effects, Proc. 14th ICOLD Congress, Rio de Janeiro.

Roberts, C. P. R. 1982. "Flow Profile Calculations." HYDRO 82, University of Pretoria, Pretoria.

Strand, R. I., and E. L. Pemberton. 1987. "Reservoir Sedimentation." In *Design of Small Dams*. Denver, CO: U.S. Bureau of Reclamation.

van Rijn, L. C. 2013. "Sedimentation of Sand and Mud in Reservoirs in Rivers." Accessed June 26, 2015. http://www.leovanrijn-sediment.com.

CHAPTER 6

Sediment Monitoring

Gregory L. Morris

Introduction

A balance between sediment inflow and outflow will occur at all reservoirs, either as a result of management or by natural processes. Absent active management, such as properly timed gate operation to pass both water and sediment, reservoirs will fill with sediment until all usable storage is lost, the delta reaches the dam, the bed load is discharged over the spillway, and the reservoir is replaced by a river, as shown in photo 6.1. *Sustainable sediment management seeks to maintain long-term reservoir capacity, retarding the rate of storage loss and eventually bringing sediment inflow and discharge into balance while maximizing usable storage capacity, hydropower production, or other benefits.* Monitoring of sediment transport, sediment deposition, and scour is required to determine the sedimentation rate and to design and effectively implement management techniques for the sustainable use of reservoirs.

Development of a sediment management program for either new or existing reservoirs will typically follow three basic steps.

Monitor and document existing or historical conditions. Historical sedimentation data are necessary to determine the rate of storage loss, identify management alternatives, and calibrate sediment transport models. This step will typically include organization and review of all available information and data on hydrology and sediment yield, updated bathymetric mapping, and sediment cores to obtain grain size and bulk density data for the deposited sediment. It may also include operation of a gauge station to collect suspended sediment data plus sampling of bed material. Monitoring may also be performed in the river reach below the dam.

Develop design and operational strategy. Design for sediment management will proceed through conceptual, preliminary, and final design phases. Conceptual design begins with the definition and more detailed evaluation of management alternatives, including sediment transport modeling of the more favorable structural alternatives and operating rules, and development of preliminary cost estimates and environmental analysis. This process serves as the basis for selecting implementation strategies. Subsequent design phases may involve more detailed

Photo 6.1 Fully Sedimented Coamo Reservoir in Puerto Rico, 1995

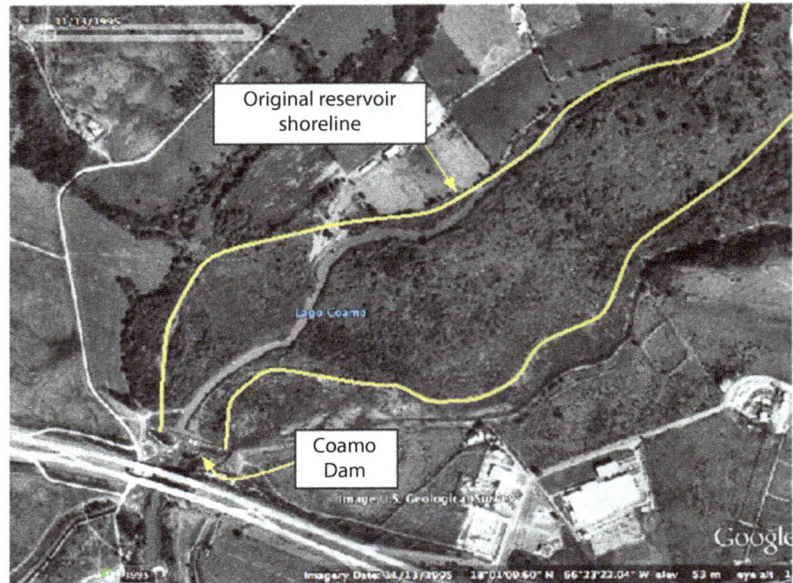

Source: U.S. Geological Survey; © Google. Further permission required for reuse.
Note: All storage volume has been lost and the dam is now a waterfall. Borrow material is now being excavated from some areas of the reservoir.

modeling, including physical modeling, and finalizing of the new operating rules and civil designs for structural measures.

Monitor effectiveness monitoring and adjust management practices. Both short- and long-term monitoring should be undertaken to provide information on the effectiveness of the implemented strategy. Short-term monitoring can be used to develop detailed information during the initial implementation period, and less intensive long-term monitoring will help determine whether any operational adjustments are required in the future.

The discussion in this chapter focuses on the monitoring of suspended sediment, which typically makes up most of the sediment that accumulates in reservoirs. Even in reservoirs with large sand deltas, the upstream river will usually transport most sand in suspension (that is, it will be measured by suspended sediment sampling and will be included in the suspended load data set).

Sampling for Suspended Sediment Load

Suspended load is the most important sediment parameter to monitor. It provides the basis for determining the expected sedimentation rate in new reservoirs and for estimating when sedimentation will begin to interfere with operations. Detailed information on the time-wise variation in sediment load is required to analyze management options that minimize the rate of deposition by sediment routing (as described in chapter 7). By measuring sediment load both upstream and downstream of a reservoir, especially during sediment release operations such as flushing, a sediment balance can be constructed to determine the effect of existing or proposed reservoir operations on the sediment balance. Data on suspended sediment concentration may also be needed to determine background environmental conditions and confirm compliance with environmental protection requirements. Detailed field and laboratory procedures for monitoring suspended sediment load are given by Nolan, Gray, and Glysson (2005) and Guy (1969), respectively.

Frequency and Duration of Sampling

Sediment transport is highly variable over time; the key to obtaining representative data is not the collection of a large number of samples, but rather to collect samples over a large range of flows. Special emphasis should be placed on sampling the infrequent high-discharge events responsible for transporting the largest portion of the sediment load. This sampling has two facets: (1) the need to sample the larger flood events that occur within any given year and (2) the need to extend the sampling period over a period of several years to increase the chances that large floods will occur and be measured within the data collection period.

Large amounts of sediment are transported by major floods, and in mountainous areas it is not uncommon for a single flood to transport more sediment than several years of "normal" flow. Figure 6.1 illustrates the impact of a large flood event at the Kulekhani reservoir in Nepal, showing that a single large event destabilized the watershed, leaving landslide scars and sediment deposits that continued to deliver large amounts of sediment into the reservoir over a three-year period. This specific pattern is not universal, however; in some watersheds an extreme event may wash out the readily mobilized sediment, resulting in reduced sediment discharge in subsequent years. Short-term data sets that do not include the impact of large sediment-producing events can seriously underestimate long-term sediment yield.

Suspended Sediment Sampling Techniques

This section provides a very brief overview of some fluvial sampling techniques. The Federal Interagency Sedimentation Project and the U.S. Geological Survey websites (http://water.usgs.gov/fisp/ and http://water.usgs.gov/osw/techniques /sediment.html) provide extensive information on many aspects of fluvial sediment monitoring and analysis. Field sampling methods are described in detail by Edwards and Glysson (1998).

The concentration of slowly settling fine particles (silts and clays) is relatively constant across a river cross-section because they are sustained in well-mixed condition by river turbulence. However, sands settle more rapidly, and strong concentration gradients can be found, with higher concentrations near the bed of the river. Consequently, a sample collected at the water surface, or at a fixed point, will not be representative of the entire cross-section. Furthermore, sand

Extending the Life of Reservoirs • http://dx.doi.org/10.1596/978-1-4648-0838-8

Figure 6.1 Loss of Storage Capacity in Kulekhani Reservoir, Nepal, Resulting from the Extreme Monsoon of 1993

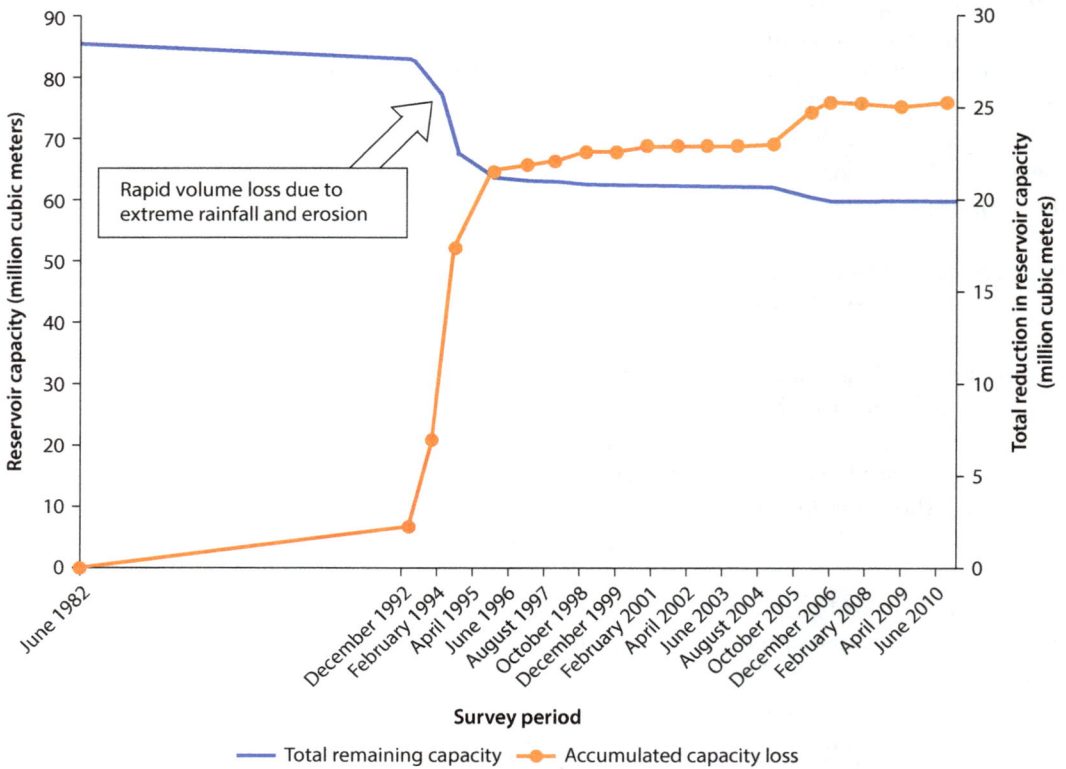

Source: Shrestha 2012.

concentration in the fast-moving water that transports most of the sediment may be higher than in the slower-moving water closer to the water's edge.

To overcome this variability, suspended sediment is commonly sampled at multiple depth-integrated verticals across a cross-section. The US D-74 sampler, suitable for use in larger streams, is illustrated in photo 6.2. It is lowered by winch from a bridge or a funicular carriage suspended by cables, or on larger rivers it may be suspended from a boat. At each vertical across the river the sampler is lowered at a constant rate until it touches the bottom, and then immediately retrieved, thereby collecting a depth-integrated sample from each vertical. This is an isokinetic sampler, so named because water enters the sampling nozzle at the same velocity as the river's surrounding flow, meaning that less water and sediment will be captured in the slow-moving portion of the river cross-section. This approach results in a discharge-weighted suspended sediment concentration, and is the most accurate way to measure sediment concentration and compute transport at any given moment. Point samplers, which sample water from discrete depths within the water column at each vertical, can also be used. The sediment samples collected should also be analyzed for particle size distribution.

Photo 6.2 US D-74 Isokinetic Suspended Sediment Sampler

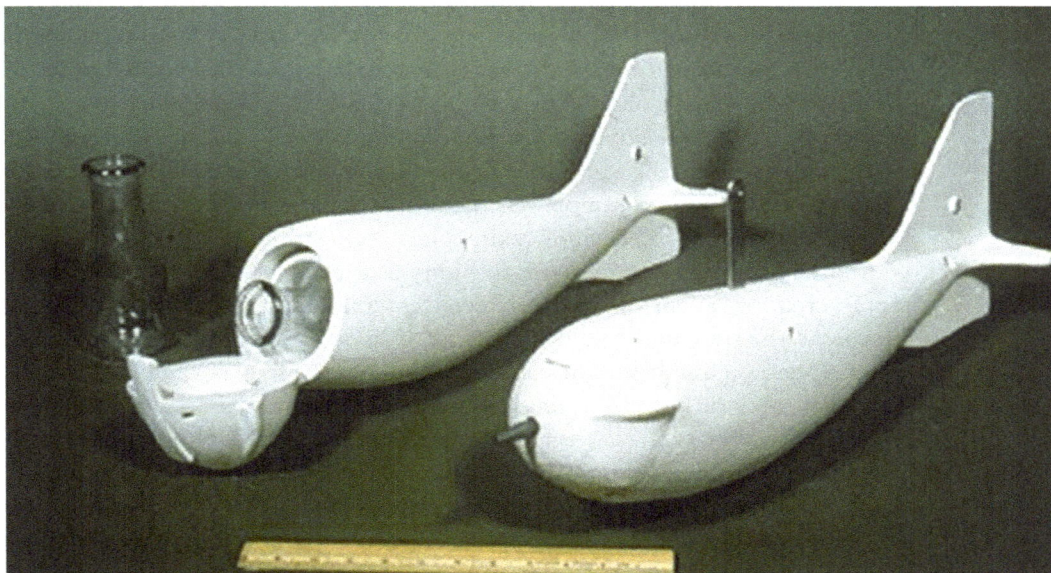

Source: U.S. Geological Survey.
Note: Ruler in foreground is 12 inches (30 centimeters) long.

However, even perfect field technique can produce a poor yield estimate if the sampling periods do not include the major sediment-transporting events. Despite designing a sampling program to target flood events, these events can be difficult or impossible to measure when they are associated with hurricanes, or when access is blocked by flooded roads and landslides. As a result, a program of depth-integrated sampling undertaken at infrequent intervals can easily miss those flood peaks that transport a disproportionate amount of the sediment load. Within a single flood event, the sediment concentration in the rising portion of the flood is frequently higher than the concentration in the falling part of the flood, making it important to measure both the rising and falling portion of the flood event. These time-wise variations in concentration, if consistent, can also be important from the sediment management standpoint. For example, the high-concentration flow at the beginning of the storm can be released downstream, while the low-concentration flow from the storm recession can be captured in the reservoir.

Several techniques have been used to capture more data during floods. Pump samplers may be installed that automatically pump water from the river into sample bottles when the river stage or turbidity rises. This method is very useful for filling data gaps, but the pump sampler obtains the concentration at a single depth at a single location, often near the river bank, which is not necessarily representative of the entire cross-section. Furthermore, a point sampler that is exposed (out of the water) during low flow may be drawing a sample from mid-depth during moderate flows, but as the flow increases and water level rises the fixed sampling point becomes deeper and deeper with respect to the full river depth. As a result, the measurement point relative to the total water depth will continually change. Suspended sands tend to be more concentrated near the bottom of the water column than at the top, and a fixed sampling point will measure at different relative locations within this vertical concentration gradient as the water level changes. For this reason, a point sampler must be calibrated against depth-integrated sampling across a full range of discharges, and a low-flow calibration cannot be simply extrapolated to high flows.

Another strategy is to measure turbidity as a surrogate for suspended sediment, creating a correlation between turbidity and suspended sediment specific to the river under study. However, turbidity is an optical characteristic of water, and on a unit weight basis, one milligram of clay per liter will create a much higher turbidity response than one milligram of sand per liter. As a result, the correlation between turbidity and suspended sediment may be quite variable, and in some instances unusable, especially if the concentration of sand relative to fines (silt and clay) changes with discharge. A work-around is to continually recalibrate the relationship by combining turbidity measurements with pumped water samples. The turbidimeter optical window can also be subject to fouling, further complicating measurement at an unattended monitoring station. Being a point-sampling technique, turbidity measurement also incorporates the inherent limitation of point sampling described previously.

Emerging sampling methods include the use of laser technology (LISST [linear, integrated solid-state tube] samplers, for example) to simultaneously measure suspended sediment concentration and grain size up to 0.5 millimeters. Acoustic methods are starting to be used as a means to measure sediment concentration, and they show considerable promise. However, these methods still have the inherent limitation imposed by point sampling, and correlation with depth-integrated sampling over a full range of flows is still recommended to calibrate the sampler result against the discharge-weighted concentration, as well as to obtain grain size data.

Sediment Rating Curves

Form of Rating Curves

Although many areas of the world now have extended streamflow data sets, suspended sediment data sets are typically rather short because sediment concentration is much more costly and difficult to measure than discharge. Therefore, it is common practice to operate a sediment gauging station for several years to define the average relationship between suspended sediment concentration and discharge over a wide range of flows. These data are used to define a *sediment rating equation* (or *sediment rating curve*) that correlates discharge with concentration (figure 6.2). Sediment data typically have considerable scatter, and the rating relationship will not be a good estimator of sediment concentration at any

Figure 6.2 Conceptual Diagram Showing Use of Two Rating Relationships

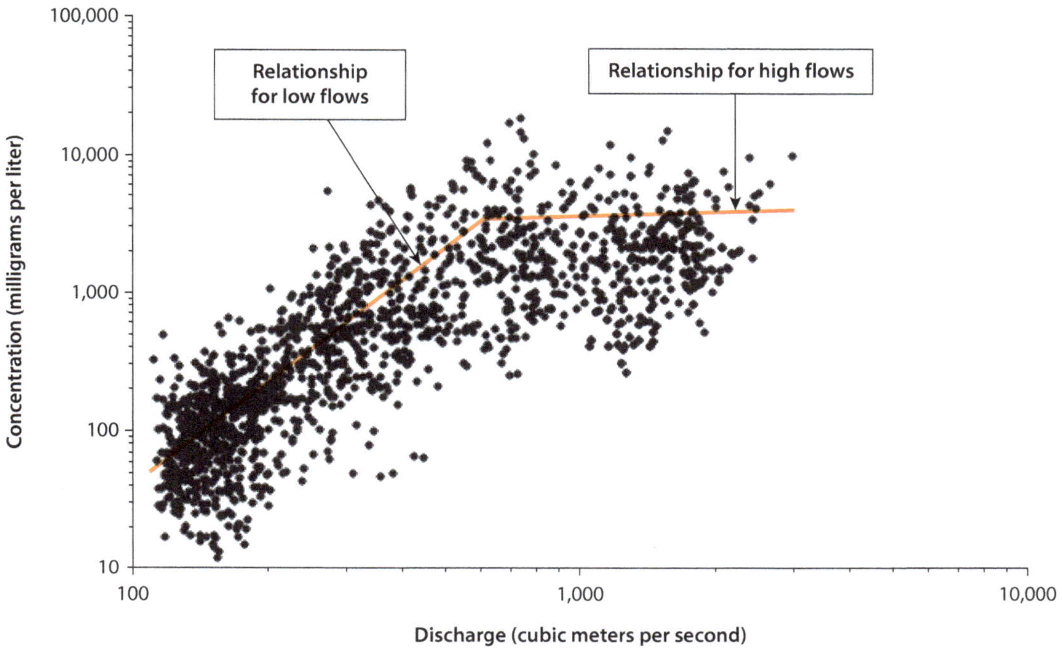

given discharge, but the relationship should be representative of the average load when applied to long-term data. Therefore, by applying this relationship to a longer daily streamflow record, the sediment load may be estimated over the entire period of record. Consult Glysson (1987) for more detailed practical considerations for constructing sediment rating curves.

Selection of Rating Equation

Suspended sediment data sets have different forms, and there is no specific mathematical equation that should be universally used. Usually the data are plotted on log-log scale, and the discharge-concentration relationship is described using a power function having the form

$$SSC = aQ^b, \tag{6.1}$$

in which SSC = suspended sediment concentration (in milligrams per liter), Q = discharge (in cubic meters per second), and a and b are coefficient values particular to each site. However, other equations may be appropriate depending on the data set. Not infrequently, multiple equations may be needed to describe the data. The application of two different equations over different flow ranges is shown in figure 6.2, and in some cases different equations may be used during different periods of the year, such as for spring snowmelt versus convective summer storms. In some cases even a linear relationship may provide the best relationship.

Extending the Life of Reservoirs • http://dx.doi.org/10.1596/978-1-4648-0838-8

Correlating Sediment Concentration with Load

Creating a relationship between discharge and sediment load, instead of concentration (SSC), is not recommended, even though it is commonly done. Fitting a regression equation to a data set using either technique will yield identical values for load. However, because the load curve, when graphed, contains discharge on both the x and y axes, it produces a spurious correlation and a higher R-squared (R^2) value than justified by the underlying data. This effect is illustrated in figure 6.3, where a concentration-discharge data set with zero correlation ($R^2 = 0$) will produce a value of $R^2 = 0.50$ by simply replotting the same data as a load-discharge relationship. If the load curve is examined, there appears to be a correlation, yet no correlation actually exists between concentration and discharge. Because discharge occurs on both axes, it produces a correlation upon itself independent of the concentration-discharge correlation in the original data set. To avoid this spurious correlation, and better visualize the data, sediment rating curves should be constructed from concentration-discharge data.

Use of a load-discharge relationship is not recommended because it makes it more difficult to understand the scatter in the underlying data and to visualize important data patterns, such as that shown in figure 6.2. Also, when extrapolating beyond the limit of the available suspended sediment data, it is important to judge whether the resulting suspended sediment concentration appears reasonable. The recommended technique is to construct a relationship between discharge and concentration, and to compute sediment load as the product of concentration and discharge.

Another error that occurs at times in suspended sediment analysis is use of the incorrect time base. In computing long-term sediment load, daily discharge data are normally used to compute load. Applying a rating equation based on instantaneous flows to mean monthly discharges, which averages out the peak discharges, will underestimate sediment load.

Figure 6.3 Spurious Correlation Caused by Incorporating Discharge on Both Graph Axes

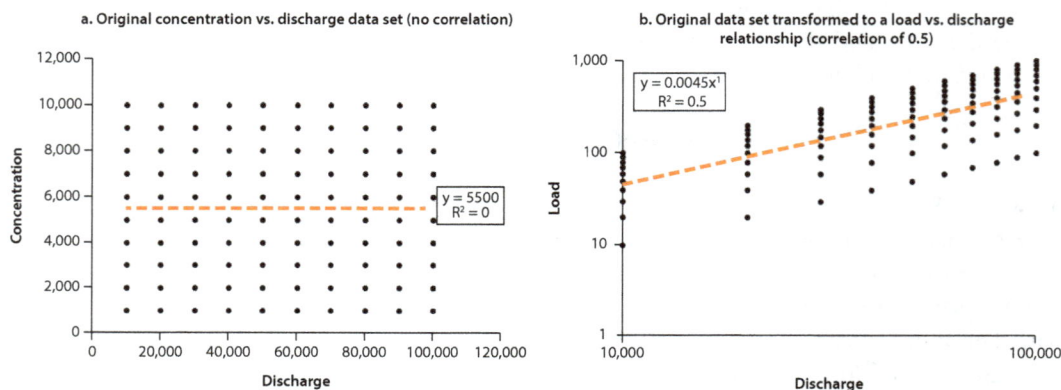

Note: A concentration-discharge data set with zero correlation (panel a) produces a correlation coefficient (R^2) value of 0.5 when the data are replotted as a load-discharge relationship (panel b). (Load = concentration × discharge.)

Curve Fitting for Rating Relationships

Most suspended sediment data sets are analyzed using a power equation, as outlined in the previous section. A potentially important source of error is the underestimation bias that occurs when a least squares regression (spreadsheet "trendline") is used to fit a power function. The degree of underestimation increases with the degree of scatter about the rating curve and can exceed 50 percent (Ferguson 1986; Asselman 2000). In the data set reproduced as figure 6.4, use of a rating equation computed by spreadsheet accounts for only 37 percent of the total load, while the nonlinear regression equation computed by a Solver spreadsheet add-in accurately reproduces the load in the original data set. Another problem occurs because a least squares regression treats all data points equally, whereas in sediment transport the data points representing higher discharge and higher concentration events are much more important given that they transport much more sediment. As noted by Glysson (1987, 15), "The statistical values computed during the regression analysis are based on the logarithmic values and therefore do not minimize the sum of the squared deviations of the actual data from the regression line." In other words, the least squares technique does not

Figure 6.4 Example of Error Introduced in Rating Equation Using a Simple Spreadsheet "Trendline" Equation

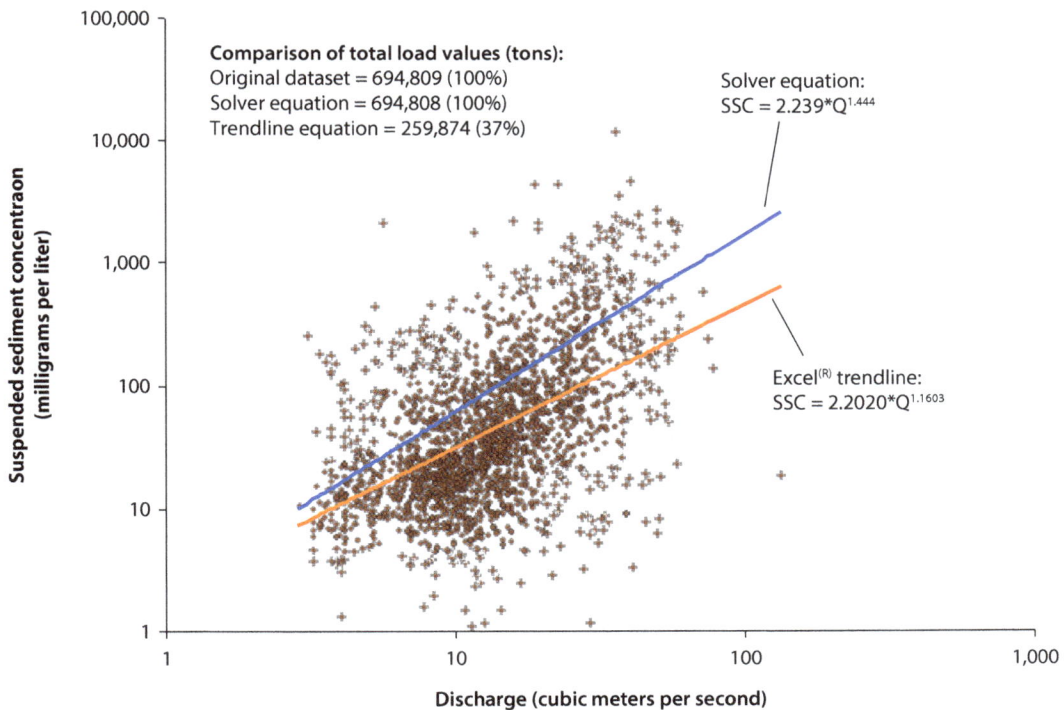

Comparison of total load values (tons):
Original dataset = 694,809 (100%)
Solver equation = 694,808 (100%)
Trendline equation = 259,874 (37%)

Solver equation:
$SSC = 2.239*Q^{1.444}$

Excel(R) trendline:
$SSC = 2.2020*Q^{1.1603}$

Note: SSC = suspended sediment concentration.

Extending the Life of Reservoirs • http://dx.doi.org/10.1596/978-1-4648-0838-8

minimize the error in the load. Note that this problem does not apply to a linear relationship.

The error that may be introduced by using a spreadsheet trendline power function to determine a sediment rating equation is also the easiest type of error to eliminate. For example, the trendline equation fit in figure 6.4 produces a load equal to only 37 percent of the true load (the true load is 270 percent greater than that estimated by the equation). Ferguson (1986) outlines a procedure to correct for the bias introduced by the regression on log-log data, but it is far from universally applicable (Walling and Webb 1988). For example, application of Ferguson's correction to this data set still only captures 73 percent of the load in the original data set, and on other data sets it can cause the load to be overestimated.

The best way to protect against serious curve-fitting errors is to back-test the equation using the original data set from which it was derived to check whether the equation can reproduce the sediment load in the original data set. This is done by applying the rating equation to each discharge value in the original data set, computing the resulting sediment concentration and load, and comparing the total load computed using the rating equation to the total measured load computed for the original data set. If this technique does not produce a reasonable match (<5 percent error), then the fitted equation should be adjusted. In figure 6.4, the adjusted equation was fitted by a nonlinear regression using an Excel "Solver" add-in, and using as an objective function the minimization of the difference between the sediment load in the original data set and that estimated by the equation.

It is also important to ensure that not only is the total load correctly reproduced but that it is reproduced over different ranges; that is, the partial load accurately reproduced for both high and low discharge ranges? If, for example, flow is to be diverted into an offstream reservoir or other facility, does the equation accurately reproduce the partial load in the discharge range that is most critical for the diverted flows? An equation that reproduces the total load over the entire data set will not necessarily produce an unbiased fit over the partial range of flows that contribute most of the diverted water or sediment. When problems fitting the equation over the entire range arise, use of two equations may be called for (recall figure 6.2).

The "correct" mathematical relationship for a rating relationship is one that best reproduces the load in the original data set over the full range of discharges. Because of the variety of shapes that rating curves can have, there is no accepted or standardized methodology. The following approach is recommended:

- Plot concentration versus discharge on different scales—arithmetic, semi-log, and log-log—visually examining the data to determine the type of equation that might be used and whether single or multiple equations should be used; delimit the appropriate data ranges for multiple equations.
- Use Excel to select the curve type and generate an initial best-fit curve.
- Compare the total and partial loads predicted by the equation with those in the original data set.

- If total load does not agree within about 5 percent of the original data set, use a "Solver" add-in to Excel to better fit the equation, using minimization of the error in load as the objective function and using the previously generated coefficient values as the starting point.
- Recheck the total and partial load results, and if the result is not satisfactory make manual adjustments until a satisfactory solution is found (for example, reconsider splitting the range into two different equations).

As a word of caution, nonlinear "solver" solutions are not unique, and the result can depend on the starting point for the solution iterations as well as the iteration algorithm used.

In some cases it may be necessary to manually adjust the equations to get the desired fit. Recall that from the standpoint of computing load, the high-discharge events are very important, and typically have little field data. Engineering judgment should be used to ensure that mathematically fitted curves do not produce unrealistic values when extended to discharges above that in the original sediment data set. Such manual adjustments may include establishing a maximum concentration value at high discharges or manually adjusting the curve to better fit the limited data at high discharges. Because high discharges are so important to sediment transport, the collection of suspended sediment data from high-discharge events and the use of judgment to properly handle high discharges in developing the rating relationship are both very important parameters for acquiring the appropriate data and developing reasonable load estimates from those data. Extreme episodic events, such as debris flows, are not captured in suspended sediment data sets and should be analyzed from a geomorphic perspective.

Short-Term Variation in Sediment Load

Sediment load is highly concentrated in time, and some sediment management methods take advantage of this variability to pass high-concentration flood events through, or bypass around, the storage to be protected against sedimentation. In many rivers, the suspended sediment concentration varies considerably over the flood period, and an evaluation of the feasibility of these pass-through or bypass techniques requires suspended sediment data covering both the rising and falling portions of the flood. Annual load data do not provide enough information to evaluate these types of sediment management strategies. In smaller watersheds large flood peaks may have a duration of several hours, thus requiring frequent data collection including use of techniques such as pump samplers or turbidity measurement to fill in the periods between manual sampling. Turbidity meters or pump samplers collect their samples from a single point in the cross-section, and their results must be correlated to depth-integrated sampling to obtain reliable data. In monsoon climates, the early monsoon may produce much higher concentrations than the later monsoon; but, with the monsoon flows extending over two or more months, daily sediment sampling may be adequate to document the time-wise change in sediment transport.

Extending the Life of Reservoirs • http://dx.doi.org/10.1596/978-1-4648-0838-8

Long-Term Trends in Sediment Load

In addition to the natural year-to-year variability of sediment yield, the sediment load entering a reservoir can also trend upward or downward over time. New roads may be opened into a previously undeveloped watershed, and the subsequent colonization by farmers can destabilize soils and dramatically increase sediment yield. This has been a common problem in developing areas. Sediment yield can also increase if the operating rule at an upstream reservoir is modified to start releasing sediment downstream or fills to the point that it traps very little sediment.

Several factors can cause sediment yield to decline. The construction of upstream reservoirs that trap sediment can dramatically decrease sediment yield downstream, as can the construction of many small storages such as farm ponds. Changes in upstream land use, such as the restoration of vegetative cover, can lead to gradual declines in sediment yield over periods of decades. This may result from active conservation measures, or by natural reforestation when marginal farms on steep slopes are abandoned as their owners migrate to the cities. In some regions, rural depopulation has resulted from military conflict or related security issues.

Bed Material Load

Bed material is the sediment normally found on the bed of the river and that, in alluvial streams, makes up its geomorphic boundary. The bed material load is the portion of the total sediment load that is transported by rolling or jumping along the bottom of a river. From the standpoint of interpreting sampling data, it may be considered to be the portion of the load that is not captured by suspended sediment sampling. It does not consist of a specific sediment grain size. For example, sand may be transported as bed load at low flow velocities and carried in suspension when the velocity and turbulence increase during high flow. In steep mountainous rivers the bed material usually consists of larger stones, such as cobbles, and sand is normally transported as suspended load.

Bed load transport occurs when flow velocities become fast enough to mobilize the bed, and obtaining representative samples of bed material under these conditions, particularly when the material in transport consists of large stones, is rarely feasible. This makes measurement of bed load transport extremely costly and difficult, especially for larger sediment sizes. For this reason, the rate of bed material transport is normally estimated by a bed material transport equation (for example, Parker 1990; Wilcock and Crowe 2003) or by assuming it makes up some percentage of the suspended load.

To estimate the bed material transport rate by equation, properly sampling the river bed grain size at multiple locations is essential, taking care to sample the coarse material and not, for example, the sand that has been deposited on top of a gravel and cobble bar, which would be transported in suspension during the flood season and measured in a suspended sediment sampling program. Global positioning system (GPS) locations and photographs of the sampling locations should be included in the data report. Because bed material transport is almost always estimated, clearly describing the computation procedure, giving the equation used or other assumptions, is important. Reservoir data may be useful for suggesting the amount of bed load transport, and the absence of a significant delta deposit of gravel or cobbles may indicate relatively low transport rates for material in that size class. However, there is no practical way to accurately estimate the cumulative bed load transport rates from reservoir sedimentation survey data.

Bed material transport is often assumed to be about 5–10 percent of the suspended load. Because it makes up a relatively small portion of the total load in most rivers, errors in estimating bed material transport will produce a relatively small error in the total load, as compared with a similar percentage of error in the suspended load. However, because all bed load is normally trapped by reservoirs and is very difficult to remove even by aggressive techniques such as reservoir flushing, it is a critical determinant of the long-term sustainability of reservoir storage.

Bathymetric Mapping of Sedimentation

Once a reservoir is in operation, sediment accumulation is measured by performing repeated bathymetric surveys. These surveys are typically performed using an echo sounder in combination with GPS equipment, continuously downloading the data to a laptop computer in the survey boat. These data are subsequently post-processed to increase the positional accuracy and remove outliers in the echo-sounding data. Bathymetric survey equipment is highly portable and can be mounted on boats as small as rubber rafts (see photo 6.3). Survey track lines may consist of a series of reservoir cross-sections, repeating lines from previous surveys, or may consist of a series of tracks designed to create a complete topographic contour map of the reservoir. Because survey speed is rather slow, typically less than 3 meters per second, a period of weeks may be required to perform a detailed survey of a large reservoir; for this reason abbreviated surveys focusing on a limited number of cross-sections or "range lines" may be used to track sedimentation in a larger reservoir once a detailed contour survey has been prepared. However, it is essential to be aware of and to correct for errors inherent in different survey methods (range-line vs. contour). Current survey techniques are summarized by Ferrari (2006) and Ferrari and Collins (2006). Reservoir survey techniques used before the advent of field-portable computers and GPS equipment are reported by Rausch and Heinemann (1968).

Bathymetric surveys have traditionally been undertaken to document the decline in storage capacity over time and determine its impact on the various reservoir pools and the reservoir's "useful life." However, when the decision is made to manage a reservoir for sustainable use instead of a limited "useful life," it becomes important to (1) reprocess existing data to extract additional information and (2) obtain additional and different types of data that are necessary to better understand the sedimentation process and to assess strategies for sustainable management. Some monitoring strategies for sustainable use are outlined by Morris (2015).

Extending the Life of Reservoirs • http://dx.doi.org/10.1596/978-1-4648-0838-8

Photo 6.3 Portable Bathymetric Equipment Used for Reservoir Surveys

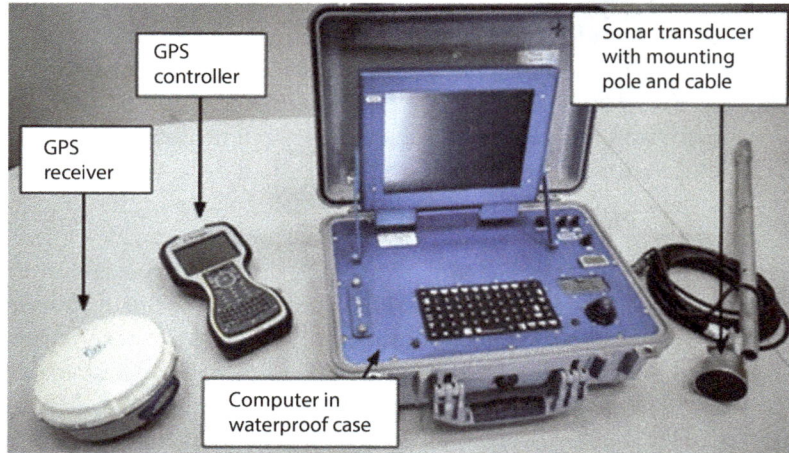

Source: © J. Portalatín. Used with permission. Further permission required for reuse.
Note: Equipment by Specialty Devices, Inc. The GPS base station antenna is not shown. GPS = global positioning system.

Reducing Bathymetric Survey Error

Bathymetric data often contain significant errors that result in uncertain sedimentation rates, as described in more detail by Morris (2015). This can be a particularly important problem during the first decades of impoundment when a smaller fraction of the total volume has been sedimented. The original pre-impoundment reservoir volume may have been computed from topographic maps, ground survey, photogrammetry, or LIDAR. However, post-impoundment surveys use bathymetric data sets, and this change in methodology can produce significantly different results without any actual change in true reservoir volume. Bathymetric surveys may be performed by measuring a series of cross-sections, from which volume may be computed by different methods, each method incorporating some degree of error. In recent years, bathymetric surveys have frequently been performed by mapping the entire reservoir instead of the traditional cross-section survey. Not only will this methodology give a different result than cross-section surveys, but the result will also depend on which software algorithm is used to compute the surface representing the reservoir bottom. For example, Ortt et al. (2000) analyze digital data sets from Loch Raven and Prettyboy reservoirs in Maryland using two different software algorithms, resulting in volume differences of 1.4 percent and 2.2 percent, respectively, due to software differences alone. With multiple potential sources of error, finding that the first bathymetric survey gives a larger reservoir volume than the pre-impoundment survey is not unusual. This situation is obviously caused by data error. However, the same type of error occurring in the other direction—causing the sedimentation rate to be overestimated—is not readily evident.

Because of these errors, it is necessary to perform repeated bathymetric surveys using a consistent methodology to obtain an accurate measurement of the sedimentation rate. Ideally, the first bathymetric survey should be performed soon after initial impoundment, and subsequent surveys and volume calculations should be performed using the same methodology. When the methodology is changed (for example, from cross-section to contour survey), the new volume should be computed by both the old and the new methodologies. Comparison of the results will show how much of the apparent volume change is due to the change in methodology and how much is attributable to sedimentation.

Following the initial bathymetric survey, additional surveys should generally be performed at intervals of no more than about 5 or 10 years until a consistent pattern of volume loss is established. Thereafter, surveys may be scheduled to correspond to a fixed percentage increment in volume loss (5 percent, for example), or following an extreme event (recall figure 6.1). The frequency of bathymetric measurement will depend on the importance of the project, with critical reservoirs with high sedimentation rates, such as Pakistan's Tarbela reservoir, being measured every year. In Colombia, bathymetric measurements are required in all hydropower reservoirs at five-year intervals.

Data Display and Interpretation

The change in reservoir volume is often plotted as a simple graph of volume versus time. Normally such a graph shows an irregular decline in volume with time because the rate of sedimentation is not constant year to year, as previously shown in the rather extreme example given in figure 6.1.

The reservoir's elevation-storage relationship is normally replotted using data from each new bathymetric survey, producing a family of curves that illustrates the change in volume over time (figure 6.5). Presenting data in this format facilitates computation of the impact of sedimentation on storage pools delimited by different elevations within the reservoir. For example, the volume change in the upper flood control pool versus the lower conservation pool may be easily tracked from these plots.

A longitudinal profile is very useful for visualizing and understanding the sedimentation processes in a reservoir. Deposits extending horizontally upstream from the dam provide clear evidence that turbid density currents are transporting significant amounts of sediment to the dam (the "muddy lake deposits" in figure 5.4). Deposits from turbid density currents consist of fine sediment that could potentially be released as they reach the dam by modifying outlet works and operational rules. However, once deposited it cannot be readily remobilized except by reservoir flushing. The longitudinal profile can also easily show the relationship between the advancing delta, operational levels, and outlet works as illustrated by the profile in figure 6.6, where the advance of the delta between 1980 and 2008 is clearly seen. Turbidity deposits are not obvious in this profile, as turbid density currents that reach the dam are released by the low-level power intake. Also, because reservoirs are typically narrow in the upstream area

Extending the Life of Reservoirs • http://dx.doi.org/10.1596/978-1-4648-0838-8

Figure 6.5 Elevation-Storage Graph: Original Relationship and Shift in Curve as a Result of Sedimentation

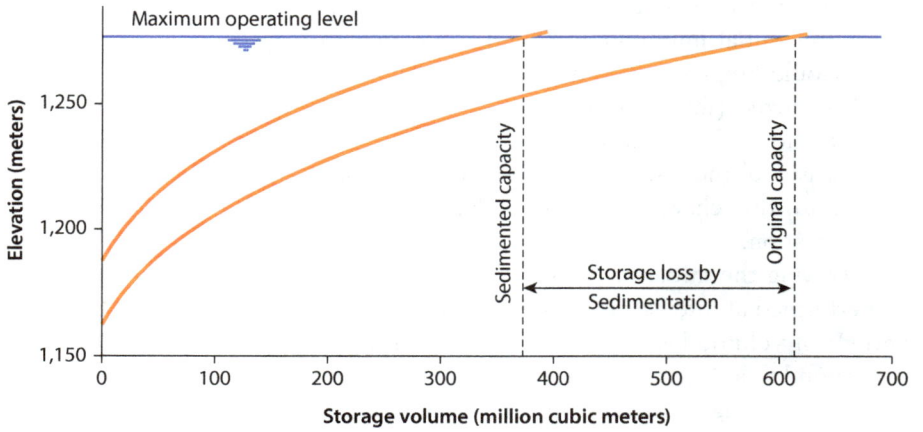

Figure 6.6 Longitudinal Thalweg Profiles of Sediment Deposits in Peligre Dam, Haiti

a. Peligre reservoir general configuration

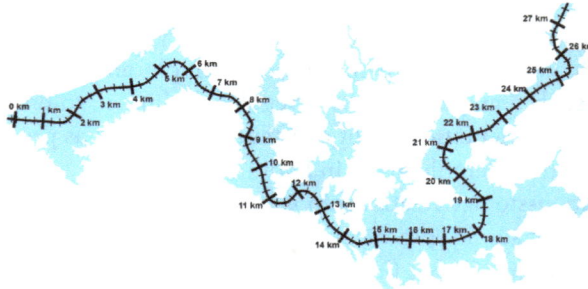

b. Peligre reservoir profile comparison

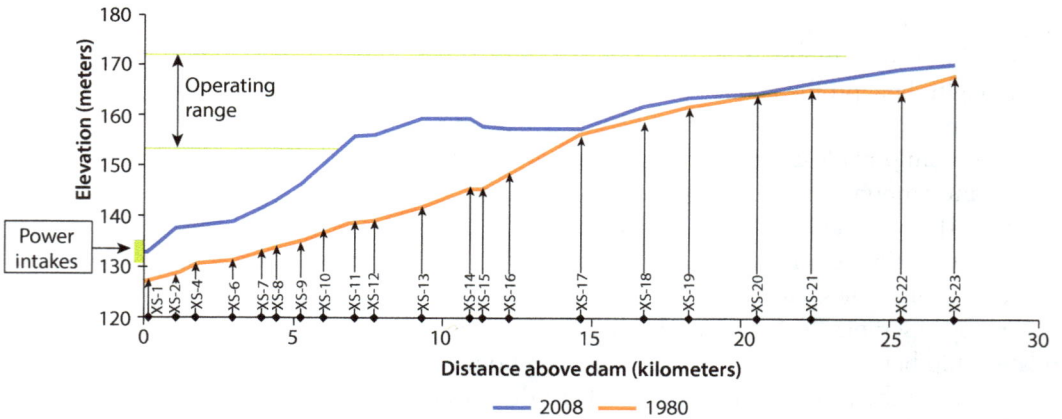

2008   1980

Source: Morris Engineering 2008.
Note: Figure illustrates longitudinal thalweg profiles of sediment deposits in Peligre Dam, Haiti, at two points in time, showing the advance of the reservoir delta toward the dam. "XS" = cross-section number.

where the delta is depositing, but widen as they approach the dam, longitudinal profiles visually exaggerate the apparent volume of sediment in the delta.

Sediment Bulk Density

The *specific weight* or *dry bulk density* is the dry weight of sediment per unit volume of deposit. Because sediment yield is expressed in terms of mass (for example, tons/year) and bathymetric surveys only measure the deposit volume, the bulk density is required to convert between sediment load and the reservoir volume displaced by sediment once it has been deposited.

Estimating Bulk Density

Bulk density depends on both the particle size distribution and the degree of compaction. The degree of compaction is influenced by whether the sediments are periodically subject to dewatering by reservoir operation (regular drawdowns that dewater the sediment make it denser), plus the degree of consolidation from self-weight and the overburden imposed by additional sediment deposits.

Values of initial (first-year) bulk density may be estimated by the Lara-Pemberton method (Strand and Pemberton 1987) based on the inflowing particle size distribution and reservoir operation. This method requires that reservoir operation be classed into one of four categories: (1) sediment always submerged or nearly submerged such that dewatering does not occur, (2) moderate to considerable drawdown during normal reservoir operation resulting in periodic dewatering of the sediment, (3) reservoir normally empty such as in a flood detention structure, and (4) riverbed sediments. The sediment composition must also be divided among the sand, silt, and clay fractions. The specific weight computation is performed by using the values in table 6.1 and equation (6.2):

$$W = W_C P_C + W_M P_M + W_S P_S, \tag{6.2}$$

in which W = is the deposit specific weight (kilograms/cubic meter); P_C, P_M, P_S = the percentages of clay, silt, and sand, respectively, for the deposited sediment; and W_C, W_M, W_S = the initial weights for clay, silt, and sand, respectively. As an example, in a continuously submerged zone within the reservoir that traps 23 percent clay, 40 percent silt, and 37 percent sand, the initial specific weight is given by equation (6.3):

$$W = (416)(0.23) + (1,120)(0.40) + (1,554)(0.37) \tag{6.3}$$
$$= 1,119 \text{ kilograms/cubic meter.}$$

Because higher elevation sediment deposits (the delta) may be composed of coarser sediment and be aerated by periodic drawdown, while deeper deposits will be continuously submerged, the reservoir should be divided into appropriate zones for a more accurate estimation of bulk density weight. More detail on this method and adjustment for compaction over time is given by Strand and Pemberton (1987) and Morris and Fan (1998).

Extending the Life of Reservoirs • http://dx.doi.org/10.1596/978-1-4648-0838-8

Table 6.1 Values of Initial Bulk Density for Use in Lara-Pemberton Equation

	Initial weight (kilograms/cubic meter)		
Operational condition	W_C	W_M	W_S
Continuously submerged	416	1,120	1,554
Periodic drawdown	561	1,140	1,554
Normally empty reservoir	641	1,150	1,554
Riverbed sediment	961	1,170	1,554

Note: W_C = weight of clay; W_M = weight of silt; W_S = weight of sand.

Sediment Sampling for Bulk Density

Sediment deposits may be sampled for bulk density, in which case a known *in situ* volume of the sample is oven dried to determine the corresponding dry weight. To obtain a representative overall bulk density it is necessary to understand the variability in sediment composition within the reservoir and obtain samples representative of that variability. Sediment is hydraulically sorted as it enters and deposits in the reservoir, meaning that sands and coarse silt will predominate upstream in the delta area, while fines are transported and deposited closer to the dam. This causes the grain size distribution and thus the bulk density to vary longitudinally within the reservoir. Sediment composition on deltas will also vary laterally; coarser material is deposited adjacent to the river channel that flows across the delta during drawdown, and finer material is deposited at the reservoir margins. The deposit density can also vary with depth because the deeper (older) deposits will have undergone more compaction. These variations need to be considered when designing a sampling program to determine the mean bulk density for the total volume of sediment trapped in the reservoir.

Fully penetrating cores at different locations along the length of the sediment deposits will most accurately represent the average sediment composition. Shallow cores can also provide useful data but require empirical adjustment for compaction. In selecting sampling locations, avoid sampling in the immediate vicinity of lateral tributaries, which can produce localized submerged deltas of coarser deposits that are not representative of the majority of the sediment in that portion of the reservoir. Take care to minimize the potential for compacting the sediment during sampling and handling.

Sediment Compaction with Time

When sand settles into a reservoir it quickly reaches its ultimate bulk density, but fine sediment compacts slowly, as illustrated in figure 6.7. During the first several years of sedimentation, the fine sediment in a reservoir accumulates on top of the original bottom and occupies a relatively large unit volume (low bulk density). As sedimentation continues, however, the underlying fine sediment deposits compact, meaning that the more recent sediment is being deposited on top of a subsiding bottom. About half of the ultimate compaction typically occurs during the first 15 years following deposition. As a result, given a constant rate of fine

Figure 6.7 Sediment Compaction over Time

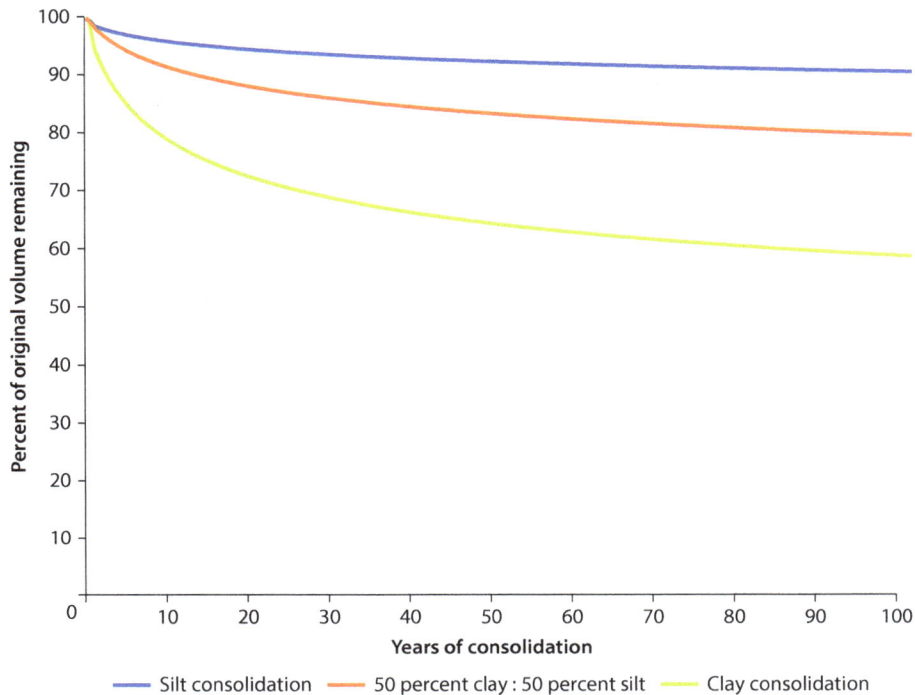

Source: Based on method of Lane and Koelzer 1943.
Note: Values are for continuously submerged deposits. Sand not shown because it does not consolidate further after year 1.

sediment deposition, successive bathymetric surveys following initial impounding will report a declining rate of volume loss, which may be erroneously interpreted as a decline in sediment yield.

Sediment Sampling of Grain Size Distribution

Sampling the grain size distribution of sediments can provide calibration information for sediment transport modeling, document the grain size transported and deposited by turbid density currents, and monitor the size of sediment approaching power intakes. However, the sampling locations must be selected to properly capture both longitudinal and horizontal variation in grain size of reservoir deposits.

The longitudinal variation in grain size is much more pronounced than the horizontal variation. Vertical variations in grain size can also be important. Sediment deposits are often layered, especially in the delta deposits, reflecting different inflow events and water levels. For example, fine sediment may be deposited downstream of the delta with the reservoir at high level, but a large inflow event that flows across the top of the delta during drawdown may scour, transport, and deposit sands on top of the previously deposited fine sediment, thereby producing a layered deposit. Also, the delta will advance over the top of older fine sediment deposits. A sample of the deposit surface will represent only the most recent sediment inflow, which may not be representative of the overall deposits, especially if the sample is taken at the end of a dry period with limited sediment inflow consisting primarily of fines.

The best way to determine the overall grain size of the deposit (and bulk densities) is by using fully penetrating cores. However, this process can be very costly and time-consuming. A rapid and much less costly approach is to use shallow cores. If sediment cores several meters deep are obtained across the entire top of the deposit, a representative "snapshot" of the sediment grain size during the recent past can be obtained, providing information for model calibration and similar purposes. Information on the rate of sediment deposition is also needed to provide an understanding of the time period that may be represented by the depth of the cores.

The upper several meters of sediment in reservoirs may be economically sampled for grain size distribution using portable vibracore equipment together with 76-millimeter (3-inch) or similar diameter tubing commonly available from irrigation suppliers. A vibracore consists of a vibrating head attached to a hollow tube; the vibrations allow the tube and its attached weights to penetrate into the sediment under self-weight. This equipment, shown in photo 6.4, can extract several cores per day, is easily transportable in a pickup truck, and is an economical method for sampling multiple locations in a short period. The penetration depth

Photo 6.4 Portable Vibracore Equipment for Sampling of Reservoir Sediments

Source: © G. Morris. Used with permission. Further permission required for reuse.
Note: Equipment by Specialty Devices, Inc.

of vibracore sampling depends on the sediment composition; penetration depths in excess of 1 meter should not be anticipated in sands, but penetration depths in excess of 3 meters can be achieved in unconsolidated silts and clays.

Summary

Sediment monitoring for sustainable management requires more detailed data collection and analysis than the approach traditionally used to merely document storage loss and reservoir "useful life." The specific collection and monitoring approaches will vary from one site to another, but should be tailored to provide the type of information needed to select the appropriate management techniques and optimize their implementation.

References

Asselman, N. E. M. 2000. "Fitting and Interpretation of Sediment Rating Curves." *Journal of Hydrology* 234 (3–4): 228–48. doi:10.1016/S0022-1694(00)00253-5.

Edwards, Thomas K., and G. Douglas Glysson. 1998. "Field Methods for Measurement of Fluvial Sediment." Book 3, Chapter C2, Techniques of Water Resources Investigations of the U.S. Geological Survey, Reston, VA.

Ferguson, R. I. 1986. "River Loads Underestimated by Rating Curves." *Water Resources Research* 22 (1): 74–76.

Ferrari, R. L. 2006. "Reconnaissance Technique for Reservoir Surveys." U.S. Bureau of Reclamation, Denver, Colorado.

———, and K. Collins. 2006. "Reservoir Survey and Data Analysis." In *Erosion and Sedimentation Manual*, edited by C. T. Yang. Denver: U.S. Bureau of Reclamation.

Glysson, G. D. 1987. *Sediment Transport Curves*. USGS Open-file Report 87–218, U.S. Geological Survey, Reston, VA.

Guy, H. P. 1969. "Laboratory Theory and Methods for Sediment Analysis." Techniques of Water-Resources Investigations of the United States Geological Survey, Book 5, Chapter C1. Washington, DC: U.S. Government. Printing Office.

Lane, E. W., and V. A. Koelzer. 1943. "Density of Sediments Deposited in Reservoirs: A Study of Methods Used in Measurement and Analysis of Sediment Loads in Streams." Report 9. Interagency Committee on Water Resources.

Morris, G. L. 2015. "Collection and Interpretation of Reservoir Data to Support Sustainable Use." SEDHYD 2015, 10th Federal Interagency Sedimentation Conference, Reno, NV, April 19–23.

———, and J. Fan. 1998. *Reservoir Sedimentation Handbook*. New York: McGraw-Hill.

Morris Engineering. 2008. "Sedimentation Study of Peligre Reservoir, Haiti." Report to Inter-American Development Bank. Washington, DC.

Nolan, K. M., J. R. Gray, and G. D. Glysson. 2005. "Introduction to Suspended-Sediment Sampling." USGS Scientific Investigations Report 2005-5077, U.S. Geological Survey, Reston, VA.

Ortt, R. A., Jr, R. T. Kerhin, D. Wells, and J. Cornwell. 2000. "Bathymetric Survey and Sedimentation Analysis of Loch Raven and Prettyboy Reservoirs." Coastal and Estuarine Geology File Report 99–4, Maryland Geological Survey.

Parker, G. 1990. "Surface-Based Bedload Transport Relation for Gravel Rivers." *J. Hyd. Res.* 28: 417–36.

Rausch, D. L., and H. G. Heinemann. 1968. "Reservoir Sedimentation Survey Methods." Research Bulletin 939, University of Missouri, Agriculture Experiment Station, Colombia, MO.

Shrestha, Hari Shankar. 2012. *Sedimentation and Sediment Handling in Himalayan Reservoirs.* Trondheim, Norway: Norwegian University of Science and Technology.

Strand, R. I., and E. L. Pemberton. 1987. "Reservoir Sedimentation." In *Design of Small Dams.* Washington, DC: Bureau of Reclamation, U.S. Department of the Interior.

Walling, D. E., and B. W. Webb. 1988. "The Reliability of Rating Curve Estimates of Suspended Sediment Yield: Some Further Comments." In *Sediment Budgets: Proceedings of the Symposium Held at Porto Alegre, Brazil, 11-15 December 1988.* IAHS Pub. 174, 337–50.

Wilcock, P. R., and J. C. Crowe. 2003. "Surface-Based Transport Model for Mixed-Size Sediment." *Journal of Hydraulic Engineering* 129 (2): 120–28. doi:10.1061/(ASCE) 0733-9429(2003)129:2(120).

CHAPTER 7

Sediment Management Techniques

Gregory L. Morris

Introduction

To sustainably develop a new dam, sediment management strategies that will be needed in the future should be identified and incorporated into the original design. Sustainable use of existing reservoirs may require changes in operating rules, structural modifications, and adaptation by users to new operating conditions. This chapter provides a basic outline of sediment management alternatives applicable to both new and existing projects. A more detailed description of sediment management options and case studies is given by Morris and Fan (1998).

Management activities to address reservoir sedimentation may be classified into four broad categories: (1) methods to reduce sediment inflow from upstream, (2) methods to pass sediment through or around the impoundment to minimize sediment trapping, (3) methods to redistribute or remove sediment deposits, or (4) methods to adapt to sedimentation. These strategies are outlined in figure 7.1, which may be used as a checklist for ensuring that all classes of strategies have been considered at a given site. A classification system based on the location in which the sediment management technique will be applied is suggested by Kantoush and Sumi (2010).

A combination of management strategies will usually be used, and the techniques most suitable for implementation will change over time as the reservoir fills with sediment. The optimum sediment management strategy may consist of a sequence of different techniques to be applied as the reservoir volume diminishes. For example, the venting of turbid density currents may initially be the only feasible technique for passing sediment through a deep and hydrologically large reservoir, but this method may no longer work and other methods may become feasible when reservoir volume has been diminished by sedimentation.

Reducing Upstream Sediment Yield

Two basic strategies may be used to reduce sediment yield entering the reservoir from the upstream watershed: (1) control soil and channel erosion at its source or (2) trap eroded sediment upstream of the reservoir. These strategies are summarized below.

Figure 7.1 Classification of Sediment Management Alternatives

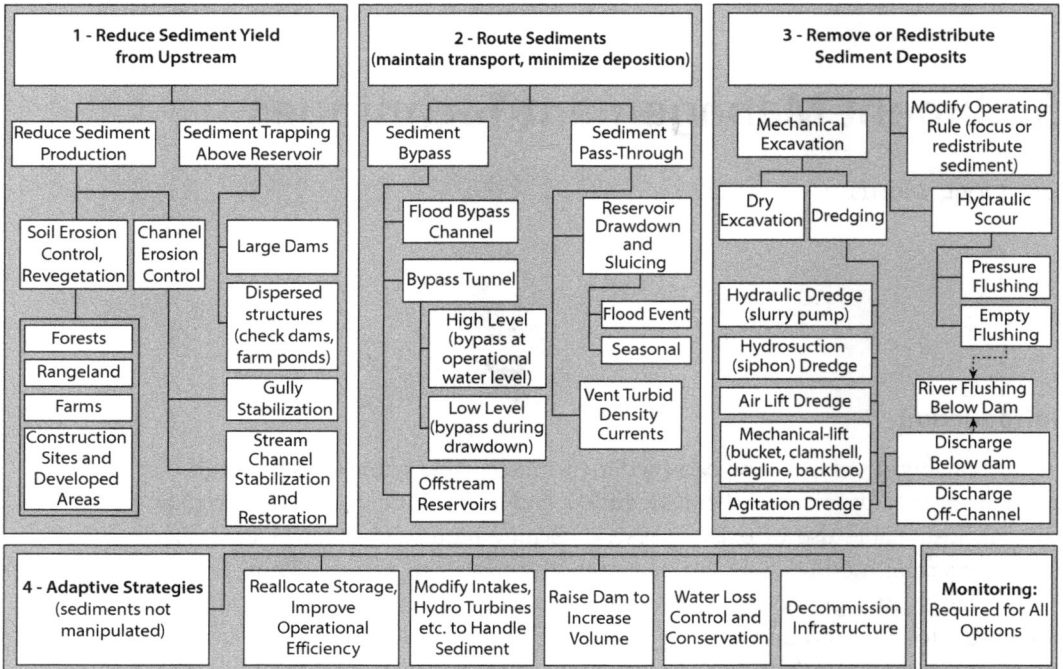

1 - Reduce Sediment Yield from Upstream

- Reduce Sediment Production
  - Soil Erosion Control, Revegetation
    - Forests
    - Rangeland
    - Farms
    - Construction Sites and Developed Areas
  - Channel Erosion Control
- Sediment Trapping Above Reservoir
  - Large Dams
  - Dispersed structures (check dams, farm ponds)
  - Gully Stabilization
  - Stream Channel Stabilization and Restoration

2 - Route Sediments (maintain transport, minimize deposition)

- Sediment Bypass
  - Flood Bypass Channel
  - Bypass Tunnel
    - High Level (bypass at operational water level)
    - Low Level (bypass during drawdown)
  - Offstream Reservoirs
- Sediment Pass-Through
  - Reservoir Drawdown and Sluicing
    - Flood Event
    - Seasonal
  - Vent Turbid Density Currents

3 - Remove or Redistribute Sediment Deposits

- Mechanical Excavation
  - Dry Excavation
  - Dredging
    - Hydraulic Dredge (slurry pump)
    - Hydrosuction (siphon) Dredge
    - Air Lift Dredge
    - Mechanical-lift (bucket, clamshell, dragline, backhoe)
    - Agitation Dredge
- Modify Operating Rule (focus or redistribute sediment)
  - Hydraulic Scour
    - Pressure Flushing
    - Empty Flushing
  - River Flushing Below Dam
  - Discharge Below dam
  - Discharge Off-Channel

4 - Adaptive Strategies (sediments not manipulated)

- Reallocate Storage, Improve Operational Efficiency
- Modify Intakes, Hydro Turbines etc. to Handle Sediment
- Raise Dam to Increase Volume
- Water Loss Control and Conservation
- Decommission Infrastructure
- **Monitoring:** Required for All Options

Source: Morris 2015.

Controlling Soil Erosion

Erosion is a natural process, but human disturbance can increase erosion rates by a factor of 10 or more. Worldwide, most of this accelerated erosion is due to agricultural activities (cropping and animal husbandry) and the resultant land degradation. Erosion control seeks to reduce the erosion rate to levels as close as possible to the natural or predisturbance rate. Erosion control activities include (1) reduction of soil surface erosion, typically by promoting or sustaining vegetative cover; (2) control of channel erosion; and (3) management of mass movement, including landslides and debris flows. An overview of erosion control techniques has been compiled by Ffolliott et al. (2013).

Nonstructural measures may be classified as either vegetative or operational. *Vegetative measures* rely on the natural regenerative properties of vegetation or the management of crop and crop residue (mulch) to protect the soil. Vegetative measures are generally less expensive than structural measures and are self-renewing once established, thereby eliminating long-term maintenance needs. If critical scour thresholds are exceeded, however, vegetation alone will not resist erosion by concentrated flows, on channel banks, for example. *Operational measures* are management and scheduling techniques that minimize erosion potential, such as organizing construction activities to minimize the area of exposed soil, or scheduling timber harvests to avoid periods of high rainfall and erosion hazard. Operational measures seek to minimize erosion rates and the need for either vegetative or structural measures.

Structural measures physically intercept the movement of water to reduce flow velocity, provide detention storage for sediment trapping, and transport flows through non-erodible structures. Structural measures may include terraces; channels that intercept and direct flow; conveyance channels lined with either vegetation or hard materials; and sediment traps including check dams, farm ponds, detention basins, and reservoirs of all sizes. Structural measures are typically expensive and require maintenance for an indefinite period. As illustrated in photo 7.1, structural measures will fail without maintenance measures, and long-term sediment trapping cannot be achieved by structures made of temporary materials.

Several key elements should be considered in developing soil erosion control strategies:

- Focus on maintaining soil to the benefit of land users. If land users (particularly farmers) see benefits such as enhanced productivity from soil conservation practices, these practices will become self-sustaining rather than dependent on costly subsidies or incentives.
- The most effective and sustainable techniques focus on maximizing vegetative and mulch cover, including the use of minimum-tillage or no-till agriculture.
- In disturbed watersheds most of the erosion comes from a small percentage of the land surface. To effectively reduce erosion, identify and focus on areas that have the highest sediment yield and are most amenable to treatment. Forested areas, for example, may contain intensely disturbed areas such as logging roads, which capture and concentrate surface runoff flows and account for a highly disproportionate amount of the total erosion (recall table 4.3). Treatment should focus on these erosion hot spots.

Photo 7.1 Gabion Check Dam, La Paz, Bolivia, Which Failed after Less Than Five Years

Source: © G. Morris. Used with permission. Further permission required for reuse.

It may take decades for the benefits of erosion control practices in the watershed to translate into reduced sediment delivery downstream. Therefore, erosion control needs to be addressed as a long-term, community-wide activity.

Gully and Channel Erosion

Land degradation and loss of vegetative cover will increase runoff, as will urbanization, which creates paved surfaces and highly efficient storm sewer networks comprising pipes and high-velocity channels. Increased runoff flow peaks can greatly accelerate erosion in downstream channels. Even without upstream disturbances, linear features such as trails and roads can capture and redirect small runoff flows to create a larger and more concentrative erosive flow. These concentrated runoff flows can form channels and greatly accelerate erosion.

The most extreme case of channel erosion is associated with gullies—erosional features characterized by a steep headcut that advances upstream, a transport zone, and a zone of deposition at the downstream end (figure 7.2). Watersheds may be affected by networks of either continuous or discontinuous gullies, and a watershed affected by a gully network may experience erosion rates more than 100 times greater than nongullied areas (Heede 1982). In deep soils with little clay, even small flows concentrated by features as small as foot trails and cattle paths can initiate extensive gullying and massive amounts of erosion. Classic work on gullying and its control was conducted by Heede (1966, 1978, 1982). Gully control guidelines have been prepared by Geyik (1986); Valentin, Poesen, and Li (2005); and Desta and Adugna (2012).

Gully formation is intimately tied to deteriorated soil and vegetation conditions in the watershed, and the overall success of control depends on treatment of the system as a whole. One gully cannot be singled out for treatment while the rest of the gully network is neglected. However, because funds are generally limited and not all gullies can be treated, selecting those sites where the highest return can be achieved at the least cost is important. Heede (1982) outlines a procedure for prioritizing gullies for treatment based on the location of the target reach within the gully network (gully order), number of tributaries, stage of gully development, and expected treatment returns across the total network. For example, using a check dam to raise the local base level in a single gully with a large number of tributaries will generally provide more benefit than treating a gully with few tributaries.

Stream channel erosion is very distinct from gully erosion and usually occurs as a gradual widening or incision of a preexisting natural channel. Bank erosion naturally occurs on the exterior of stream meanders, but tends to be offset by sediment deposition onto the point bar located on the opposite (interior) side of the bend, as illustrated in figure 7.3. Bank erosion is not a source of accelerated sediment yield unless the channel is increasing its dimension (widening or incising). Channel erosion control tends to be both difficult and costly, and control techniques must be tailored to each particular stream. Furthermore, it is not merely the channel itself that needs consideration, but also the stream corridor, which includes the channel banks and adjacent floodplain. The importance

Figure 7.2 Conceptual Longitudinal Profile of Gully Erosion

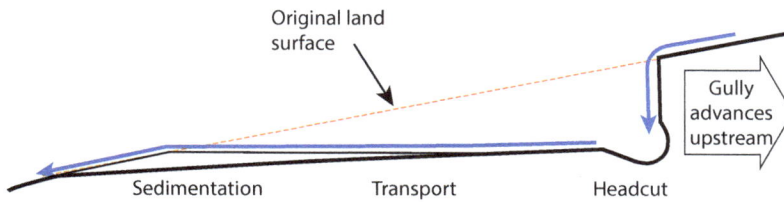

Original land surface

Gully advances upstream

Sedimentation   Transport   Headcut

Source: Adapted from Leopold, Wolman, and Miller 1964.

Figure 7.3 Lateral Migration of a Natural Stream Channel

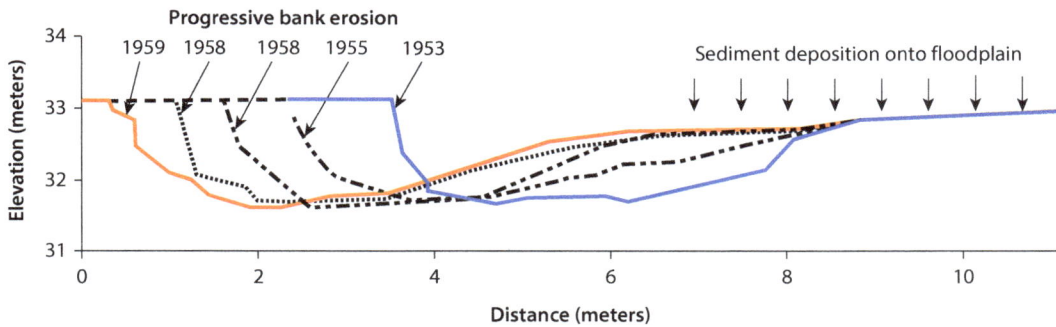

Progressive bank erosion

1959 1958 1958 1955 1953

Sediment deposition onto floodplain

Elevation (meters)

Distance (meters)

Source: Adapted from Leopold, Wolman, and Miller 1964.

of bank vegetation in aiding stream stability should also be considered. For long-term control of channel erosion, it is important that control methods help the stream move toward a naturally stable condition. Natural channel restoration principles are outlined by Rosgen (1994, 1996), and comprehensive stream channel restoration guidelines have been published by the U.S. Federal Interagency Stream Restoration Working Group (1998).

Mass Movement

Landslides and debris flows are special cases of erosion on a massive scale, with the potential for sudden and catastrophic damage. Dams and reservoirs can be severely damaged or even destroyed by mass movement, such as the 1963 landslide at the 260-meter tall Vaiont (or Vajont) arch dam in Italy, described in multiple websites. Geologic assessment should be undertaken to identify and avoid hazard areas. Activities that can destabilize marginal slopes include deforestation, road cuts or quarries that undermine the slope, and any activity that increases moisture in the unstable material. Roads on unstable slopes should generally maximize construction on fill rather than cut into the slope, and should ensure rapid drainage, including construction of drains to prevent water from entering any tension cracks that may form. On landslide-prone slopes, often the most important preventive action is to divert water away from the unstable slope and plant trees to both stabilize soil and help withdraw moisture. However, when unstable slopes receive excess moisture, mass movement may be inevitable.

Extending the Life of Reservoirs • http://dx.doi.org/10.1596/978-1-4648-0838-8

Geotechnical solutions are expensive and feasible only when costly infrastructure needs to be protected. An overview of techniques for addressing landslide problems is given by Marui (1988).

Debris flows can be controlled by permeable sabo dams and properly designed conveyance channels. Sabo dams are designed to allow clear-water floods to pass, but offer enough resistance to trap debris flows. They have been used extensively in Japan (Chanson 2004; Kantoush and Sumi 2010), but may be applied in a wide variety of environments. Photo 7.2 shows a sabo dam constructed using steel piles in a stream above Quito, Ecuador, to control debris flows resulting from ash deposits by active volcanoes.

Upstream Sediment Trapping

Not all eroded sediment will reach a channel or a downstream reservoir. Sediment trapping naturally occurs in small soil depressions and behind small obstructions, and even sediment delivered into a stream can be redeposited within the channel or on the floodplain when the river overspills its banks. Because of this natural trapping, the volume of sediment delivered to reservoirs is normally significantly less than the amount of erosion. The ratio of sediment eroded to sediment delivered is termed the *sediment delivery ratio*. It is difficult to convert erosion estimates to a reliable estimate of sediment yield because of the high degree of uncertainty in estimating the sediment delivery ratio (Walling 1983, de Vente et al. 2007).

Reservoir construction is the most reliable method for reducing sediment yield downstream. Sediment trapping occurs in impoundments of all sizes, ranging from large storage reservoirs to small check dams and farm ponds. The combined effect of numerous small dispersed structures can be large. For example, there are at least 2.6 million small farm ponds that capture runoff from 21 percent of the total drainage area of the conterminous United States, representing

Photo 7.2 Sabo Dam above the City of Quito, Ecuador

Source: © G. Morris. Used with permission. Further permission required for reuse.

25 percent of total sheet and rill erosion (Renwick et al. 2005). However, if not built of erosion-resistant material and properly constructed and maintained, sediment storage structures will eventually be breached and release stored sediment, as was illustrated in photo 7.1. In a review of 70 check dams installed in a semi-arid watershed in the southwestern United States, Gellis et al. (1995) found that 47 percent of the structures had failed and several more were near failing after about 50 years. Reasons for breaching included internal erosion (piping), downstream scour, undersized spillway, and active arroyo deepening and widening. It was noted that check dams are most effective in reducing sediment yield when used to stabilize conditions to the point that revegetation can occur and the gullying process be arrested.

Large dams are highly effective sediment traps until they become filled with sediment or begin operations to release sediment. Their long-term sediment trapping efficiency can be estimated from the Brune curve (see figure 5.1). The construction of large upstream impoundments for the sole purpose of trapping sediment has rarely been found to be a cost-effective sediment management technique. However, sediment retention ponds installed immediately downstream of activities that generate high rates of sediment discharge, such as construction sites or mining operations, are the most effective way to trap sediment from these highly disturbed erosion hot spots.

Sediment Routing

Sediment routing refers to a family of techniques that take advantage of the time-wise variation in sediment discharge, managing flows during periods of highest sediment yield to minimize sediment trapping in the reservoir. *Sediment bypass strategies* include (1) diverting clear water into a reservoir while selectively excluding sediment-laden flood flows and (2) bypassing sediment-laden flood flows around an onstream reservoir. *Sediment pass-through strategies* include (1) reservoir drawdown to pass sediment-laden floods through the impoundment at a high velocity to minimize deposition, termed *sluicing*, and (2) venting of turbid density currents through a low-level outlet. In all cases the objective is to release sediment-laden water and impound clear water.

Sediment routing techniques require that a fraction of the river inflow be dedicated to transporting sediment around or through the reservoir, and is not feasible if all the inflow is being captured and stored. However, as reservoir capacity is diminished by sedimentation, sediment routing strategies become increasingly feasible. Therefore, it may become a viable future strategy at sites where it is not currently feasible.

Sediment Bypass

Sediment can be bypassed by constructing an *offstream reservoir* outside the natural river channel, either by impounding a side tributary or by constructing the impoundment on an upland area. Clear water is diverted into the offstream reservoir by a river intake, but large sediment-laden floods are passed beyond the

Figure 7.4 Basic Features of Conventional Onstream Reservoir Compared with Offstream Reservoir

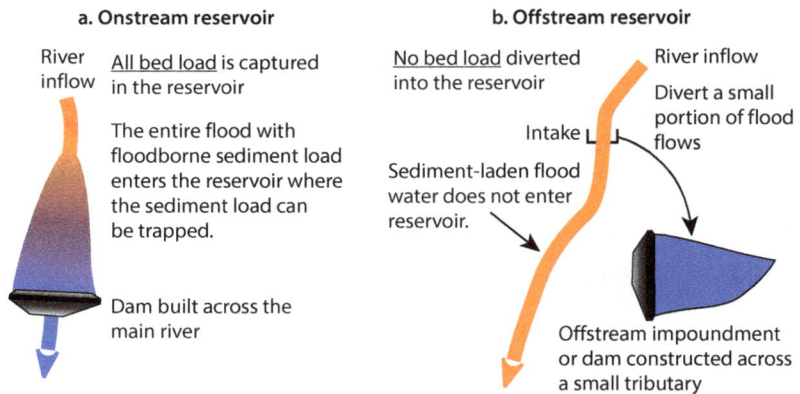

a. Onstream reservoir

River inflow

All bed load is captured in the reservoir

The entire flood with floodborne sediment load enters the reservoir where the sediment load can be trapped.

Dam built across the main river

b. Offstream reservoir

No bed load diverted into the reservoir

River inflow

Divert a small portion of flood flows

Intake

Sediment-laden flood water does not enter reservoir.

Offstream impoundment or dam constructed across a small tributary

intake and are not diverted into storage, as shown in figure 7.4, which contrasts onstream and offstream reservoirs. Offstream reservoirs have been used for water supply reservoirs and for daily regulation storage in run-of-river and pumped storage schemes. Although highly effective in reducing sedimentation, provision should be made for the eventual cleanout of offstream reservoirs.

Sediment enters an offstream reservoir either as suspended inflow from the diverted stream or by erosion from the watershed tributary to the dam. Simulations for the gravity-fed Río Fajardo offstream reservoir in Puerto Rico show that 26 percent of the total streamflow can be diverted into the reservoir with only 6 percent of the suspended sediment load. Additionally, the intake design excludes 100 percent of the bed material load. However, sediment eroding from the small watershed tributary to the dam will be trapped with essentially 100 percent efficiency, since the reservoir is operated to avoid spills. For this reason, in developing offstream reservoir sites it is important to minimize the catchment area above the dam and to undertake strict land use controls or convert to permanent forest to minimize long-term sediment yield (Morris 2010).

Sediment Bypass Tunnel or Channel

Discharge events that transport high sediment loads can be bypassed around an instream reservoir using a high-capacity channel or a tunnel, as illustrated in figure 7.5. Because of the unique site characteristics required for construction of a bypass channel and the high cost of tunneling, sediment bypass tunnels have been used in only a few hydropower reservoirs in mountainous areas of Japan and Switzerland, but they can be useful for retrofitting dams that were not constructed with outlets for sediment management. Examples are provided by Sumi, Okano, and Takata (2004); and Sumi and Kantoush (2010). A retrofit sediment bypass tunnel at the Solis Dam in Switzerland was described by Auel, Berchtold, and Boes (2010) and Auel et al. (2011). Interest in this technology is increasing as sedimentation problems become more pronounced.

Figure 7.5 Alternatives for Bypass of Sediment-Laden Floods

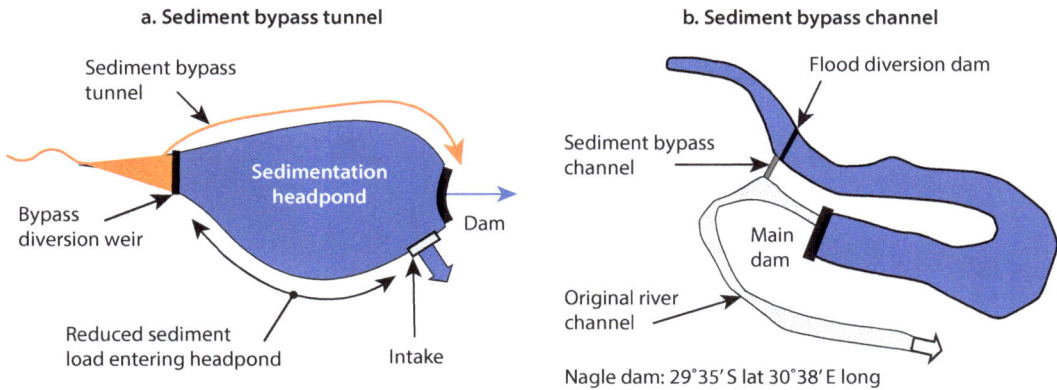

a. Sediment bypass tunnel

Sediment bypass tunnel

Sedimentation headpond

Bypass diversion weir

Dam

Reduced sediment load entering headpond

Intake

b. Sediment bypass channel

Flood diversion dam

Sediment bypass channel

Main dam

Original river channel

Nagle dam: 29°35′ S lat 30°38′ E long

The entrance of a sediment bypass tunnel is set upstream of the area that will be protected from sedimentation, and diverts either suspended load or both bed and suspended load to a discharge point below the dam. A high-level bypass tunnel is located upstream of the reservoir and can bypass bed material without reservoir drawdown, as shown in panel a of figure 7.5.

At longer reservoirs where it is not feasible to extend the tunnel to the upstream end of the impoundment, a low-level tunnel may be placed closer to the dam to bypass suspended sediment during periods of power production to reduce sediment load entering the intake area. However, a low-level tunnel near the dam cannot bypass bed material without reservoir drawdown. Sediment trapped in the headpond area downstream of the tunnel entrance must be removed by flushing or other methods.

Tunnel flow velocities typically exceed 10 meters per second, and the bypass tunnel floor is lined with abrasion-resistant material including high-strength concrete, granite blocks, cast basalt tiles, and steel plate. Currently, the longest bypass tunnel is the 4.3-kilometer, 300-cubic-meters-per-second, low-level tunnel at the Miwa Dam in Japan. A variety of projects involving sediment bypass tunnels are described in Boes (2015).

The largest-capacity bypass tunnel to date is at the Nanhua (Nan-Hwa) water supply reservoir in Taiwan, China. A bypass tunnel was originally proposed (as described in Morris and Fan 1998), but was not included in the construction budget. However, after losing 37 percent of its capacity to sedimentation, and two typhoons that approached the spillway's capacity, construction of a 1,000-cubic-meter-per-second low-level bypass tunnel began 21 years after initial impounding. This tunnel is expected to enter operation in 2018 and have a bypass efficiency for suspended sediment of 24 percent, reducing sedimentation as well as increasing flood discharge capacity (Kung et al. 2015).

At run-of-river hydropower dams a bypass tunnel can be used to create a sediment trap in the headpond area, reducing both the sediment and hydraulic

Table 7.1 Operational Strategy for Sediment Bypass Tunnel at Run-of-River Hydropower Dam, Using Sedimentation Headpond Instead of Desanding Basin

Reservoir inflow	Operation
Inflow < design power flow	All inflow to power
Design power flow < inflow < flushing flow	Inflow exceeding design power flow is bypassed
Inflow > flushing flow threshold	Gates at dam opened, empty flushing of reservoir

load in the headpond during power production (panel a of figure 7.5). Sediment is periodically removed from the headpond by flushing. Using an operating rule similar to that in table 7.1, this strategy may eliminate the requirement for desanding basins, an especially attractive option if costly underground basins would otherwise be required.

Because bypass tunnels normally operate for extended or multiple periods each year, transporting sediment downstream during repeated events, they have lower environmental impacts than the larger and more concentrated sediment releases characteristic of reservoir emptying and flushing using low-level outlets at the dam. Environmental flows may be released through the bypass tunnel or at the dam, depending on project characteristics.

Sediment Pass-Through by Drawdown

Sediment can be passed through the reservoir by *sluicing*, which involves drawing down the water level during periods of high discharge and sediment load, thereby increasing flow velocity while reducing both residence time and sediment trapping. Aggressive drawdown, reducing the reservoir level to create riverine flow along the impounded reach during a flood, may also scour and release a portion of the previously deposited sediment. Drawdown sluicing is performed during the monsoon in some Himalayan run-of-river hydropower dams to preserve peaking capacity. Drawdown may also be timed to match individual floods, guided by real-time reporting gauges and hydrologic modeling, as schematically illustrated in figure 7.6 for a storage reservoir. In this case, a real-time hydrologic model is needed to predict inflow volume and guide gate operation to pass sediment-laden floods through the reservoir by the following sequence of actions.

- *Normal operation.* The system continuously monitors rainfall and streamflow data to maintain up-to-date soil moisture balance computations (antecedent moisture) in the model. Reservoir gates remain closed, or are opened to pass small floods.
- *Begin drawdown.* As a major weather system approaches and begins to produce significant rainfall in the watershed, the real-time hydrologic model continuously predicts the total volume of inflow that will enter the reservoir during the next 24 hours. When the model predicts an increasing volume of inflow that has not yet reached the reservoir, the gates are progressively opened to release water from the reservoir, matching the released

Figure 7.6 Sediment Sluicing in a Storage Reservoir during a Short-Duration Flood Event

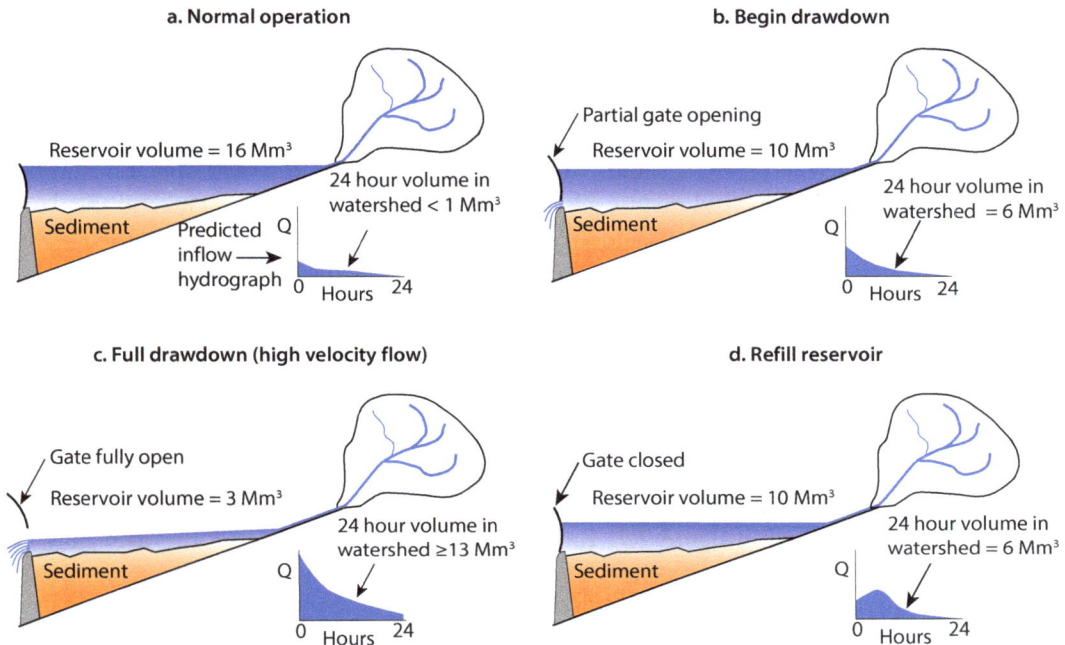

a. Normal operation

Reservoir volume = 16 Mm³

Sediment   Predicted inflow → hydrograph

24 hour volume in watershed < 1 Mm³

Q

0   Hours   24

b. Begin drawdown

Partial gate opening

Reservoir volume = 10 Mm³

Sediment

24 hour volume in watershed = 6 Mm³

Q

0   Hours   24

c. Full drawdown (high velocity flow)

Gate fully open

Reservoir volume = 3 Mm³

Sediment

24 hour volume in watershed ≥13 Mm³

Q

0   Hours   24

d. Refill reservoir

Gate closed

Reservoir volume = 10 Mm³

Sediment

24 hour volume in watershed = 6 Mm³

Q

0   Hours   24

Source: Morris and Fan 1998.
Note: Mm³ = million cubic meters.

volume to the volume predicted to enter the reservoir during the next 24 hours. The drawdown rate is limited by the critical discharge rate that begins to produce downstream flooding. Once the discharge is reached that will initiate downstream flooding, outflow should be limited to the maximum inflow rate to prevent an increase in downstream flood levels. Runoff prediction should be based on antecedent moisture and the measured rainfall received (rather than predicted rainfall), with continuous validation of predicted hydrology by real-time data from stream gauges.

- *Full drawdown.* With the reservoir fully drawn down, water flows through it at the maximum velocity possible, as in river flow. Volume in the reservoir plus volume predicted to enter the reservoir during the next 24 hours are continuously calculated to ensure that the total water volume upstream of the dam always exceeds the total reservoir storage capacity, thereby ensuring it can be completely refilled at the end of the flood.

- *Refill.* As rainfall decreases, the hydrologic model will identify the rate of gate closing necessary to ensure the reservoir will totally refill during the next 24 hours.

The real-time hydrologic monitoring and prediction system needed to implement this type of operation can be readily implemented given today's data collection platforms and hydrologic modeling techniques. Because sluicing passes the natural hydrograph and sediment load through the reservoir, natural flood flows are maintained below the dam and downstream environmental impacts are minimized.

Sediment sluicing requires large gates. When a dam lacks the required gate capacity, retrofitting the structure to install high-capacity gates may be feasible. As an example, the 80-year-old cascade along Japan's Mimikawa River is being modified to facilitate sediment sluicing to sustain hydropower operations in the face of extreme sediment loads resulting from typhoon floods. Three of the dams will have their spillways notched and large-capacity crest gates installed, and the operating rule will be modified to incorporate sediment sluicing two or three times per year along the entire cascade. The remaining two dams have sufficient gate capacity and structural modifications are not required. (Sumi et al. 2015)

Sediment Pass-Through by Turbid Density Current

Sediment-laden water is denser than clear water. With sufficient suspended sediment concentration, the inflowing turbid water will plunge beneath the clear water and flow as a turbid density current along the floor of the reservoir, as illustrated in figure 7.7. Turbid density currents that reach the dam may be released if a low-level outlet operates continuously during flood events (as would normally occur at hydropower plants), or if the low-level outlet is operated based on the predicted or monitored arrival time of the turbid density current at the dam. The efficient release of these density currents depends on successfully predicting the arrival time at the dam and operating outlets to minimize the settling period in the muddy lake. If turbid density currents reach the dam and are not released, they will deposit horizontal sediment beds (wedge deposits) extending upstream from the dam, the result of sedimentation

Figure 7.7 Passage of a Turbid Density Current through a Reservoir

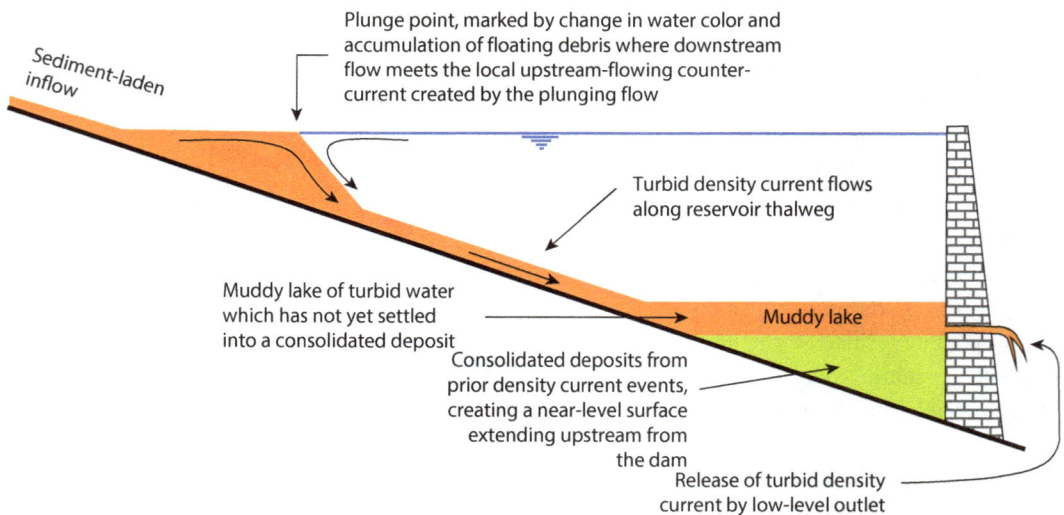

Sediment-laden inflow

Plunge point, marked by change in water color and accumulation of floating debris where downstream flow meets the local upstream-flowing counter-current created by the plunging flow

Turbid density current flows along reservoir thalweg

Muddy lake of turbid water which has not yet settled into a consolidated deposit

Muddy lake

Consolidated deposits from prior density current events, creating a near-level surface extending upstream from the dam

Release of turbid density current by low-level outlet

Source: Morris and Fan 1998.

from the muddy lake. Figure 7.7 illustrates key features of a turbidity current flowing through a reservoir.

Because turbid density currents do not require drawdown of the reservoir or similar operational measures, they can be well suited for sediment release starting in the first years of reservoir operation. Hydropower facilities with low-level power intakes may be well suited for releasing turbid density currents as long as only fine sediment reaches the dam. In hydropower plants the mechanical designers want to minimize the potential for abrasion damage to equipment by sediment, and may prefer that the sediment be trapped in the reservoir instead of passed through the turbines. However, because fine sediment comprises the great majority of sediment trapped in most reservoirs, releasing fine sediment can significantly retard the rate of volume loss and delay the delta's arrival at the area of the power intake. Reservoir deltas normally contain highly abrasive coarse sediment that cannot be passed through turbines without causing extreme damage. Thus, from the standpoint of sustainability, the release of fine sediment can significantly extend project operation at the cost of a modest increase in turbine abrasion during the initial decades of operation.

Multilevel or selective-withdrawal outlets are a standard feature at many dams for water quality management and are used to selectively withdraw or mix water of different temperatures and depths to meet downstream water quality requirements. This same approach can also be used to release deep turbid density currents in deep reservoirs where it may not be practical to install a high-pressure low-level gate. A multilevel intake or a *turbidity siphon* may be installed to aspirate turbid water from deeper levels in the reservoir for discharge through a higher level outlet. Figure 7.8 illustrates turbidity siphon configurations for releasing deep water through a higher-level outlet. A turbidity siphon of the type illustrated in panel a of figure 7.8 is currently under construction at the Zengwen reservoir in southern Taiwan, China, and the ungated curtain-wall configuration for the release of turbid flood water of the type shown in panel b of figure 7.8 has been installed at the Katagiri Dam in Japan.

Figure 7.8 Turbidity Siphon Configurations for Releasing Turbid Density Currents

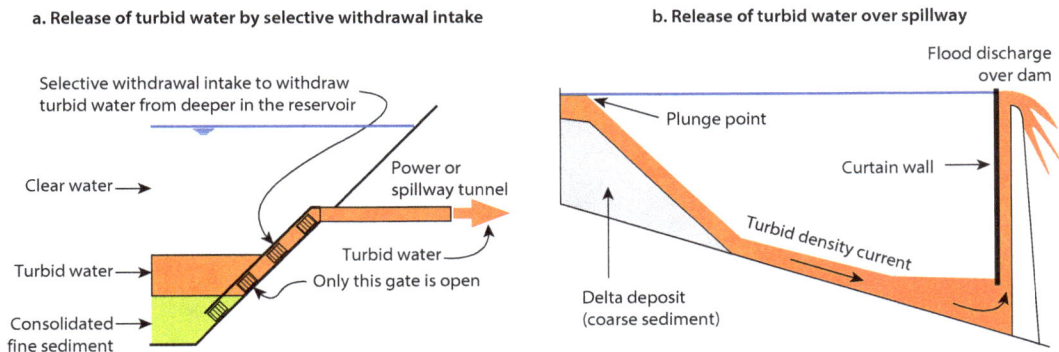

a. Release of turbid water by selective withdrawal intake

b. Release of turbid water over spillway

Redistributing or Removing Sediment Deposits

Modifying Minimum Operating Level

The principal technique used to modify the sedimentation pattern, or to redistribute deposited sediment, is to manipulate water levels in the reservoir. Reservoir deltas are normally composed of coarse sediment that cannot be passed through turbines without causing severe damage. Every time the reservoir is drawn down the river flows across the top of the delta and scours sediment, moving it downstream and closer to the power intake. This downstream progression of the delta is clearly evident at Peligre reservoir (recall figure 6.6). To slow the advance of the delta, the reservoir's minimum operating level may be gradually raised, focusing delta deposition into the upper portion of the reservoir. Figure 7.9 compares delta advancement for a constant minimum operational level against an increasing minimum operational level, showing that by gradually increasing the minimum operating level the downstream advance of the delta is retarded. As a trade-off, raising the minimum operating level accelerates the decline of operational storage volume.

Sediment Removal by Dredging

Dredging refers to any system used to remove sediment from beneath the water. *Hydraulic dredging* of submerged sediments can remove sediment from reservoirs without requiring the reservoir to be emptied, and a slurry pipeline is a clean and efficient method of transporting this sediment. The principal components of a hydraulic dredging system discharging into a containment area are shown in figure 7.10. A basic primer on dredging technology is presented by Turner (1996).

The availability of land for the disposal of dredged material is an important limitation to sustaining long-term reservoir capacity by dredging. In some instances it is permissible to discharge dredged material to the river channel downstream of the dam. Discharge below the dam is advantageous in that it

Figure 7.9 Delta Advance Depending on Reservoir Operational Levels

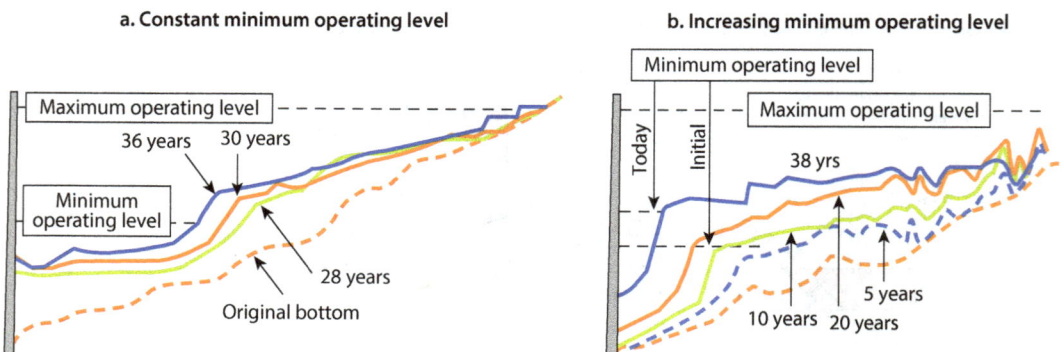

a. Constant minimum operating level

b. Increasing minimum operating level

Source: Morris 2015.

Figure 7.10 Schematic of Dredging System Components

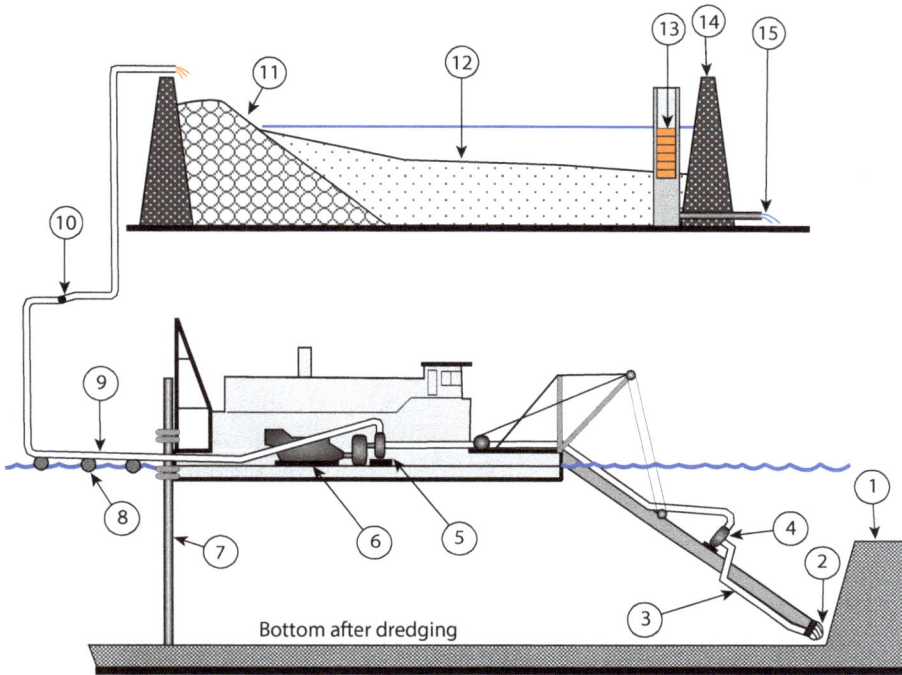

Source: Adapted from Morris and Fan 1998.
Note: The components shown in the figure are (1) sediment to be dredged; (2) rotating cutterhead to cut and suspend sediment; (3) suction line connected to the ladder; (4) ladder pump; (5) main pump; (6) main drive, either diesel or electric; (7) spud, which serves as an anchor for pivoting the dredge; (8) pontoons to support discharge pipe; (9) discharge pipeline; (10) booster pump; (11) coarser material deposited near the discharge point; (12) fine sediment; (13) discharge weir with flashboards to allow elevation to be raised as the containment area is filled; (14) containment area dike; and (15) discharge of clarified water back to the reservoir or to other receiving body.

sustains the flow of sediment along the channel. However, dredged sediment is released continuously rather than being timed to coincide with natural discharge events. Dredging may remove many years of sediment deposits in a single year, and if dredging occurs near the dam the discharged sediment will not contain the coarse material contained in the delta and needed to restore the channel below the dam. The feasibility of discharging dredged material below the dam will depend on how effectively these issues can be addressed. Nevertheless, at smaller dams in mountainous areas with frequent downstream releases, and with most of the dredged material consisting of coarse sediment, discharge below the dam can be a good alternative if sediment can be temporarily stored in-channel and then scoured and mobilized downstream by natural flood events, or if dredging is performed either continuously or annually. When downstream sediment discharge is not feasible, dredging can be sustained only as long as sufficient space is available in containment areas close enough for economical slurry pumping.

Dredging is inherently costly. It requires pumping slurry containing both water and sediment, and a slurry pipeline must be designed to transport the largest grain sizes in the material to be dredged. The high velocity required to sustain sand or coarser material in suspension generates high friction loss, which requires high energy input for pumping, and also abrades the pipeline. Slurry velocity, pumping costs, and abrasion damage are lowest when removing uniformly fine-grained material. Although a variety of novel dredging systems exist, including automated systems, all of them require pumping energy to move the dredged slurry and are subject to abrasion, and even novel systems cannot escape these major cost items.

Dredging is typically much more costly than creating storage volume during dam construction. While costs vary regionally, Allen and Dunbar (2005) estimate that recovering reservoir volume in Texas (United States) by dredging was more than double the cost of new dams. Dredging costs include engineering and permitting, acquisition and management of the dredged material placement site, and the cost of dredging itself. Under current conditions (as of 2016) reservoir dredging will typically cost more than US$3 per cubic meter of sediment removed, not including engineering, permitting, and disposal site or pumping distances more than 2 kilometers. Use of electric drives on the dredge can significantly reduce energy costs, especially at hydropower dams that can self-supply the electricity.

Hydrosuction dredging (also called siphon dredging) is a special case of hydraulic dredging that does not use a pump (Hotchkiss and Huang 1995). Instead, the motive force for transporting slurry through the pipeline is provided by the elevation differential between the reservoir level and the foot of the dam where the slurry pipeline discharges. Because the maximum energy available for slurry transport is limited by the dam height, operation of a hydrosuction dredge will typically be limited to within a few kilometers of the dam.

Dry Excavation

In some instances dry excavation has been used for sediment removal, as illustrated in photo 7.3. Unlike dredging, it requires that the reservoir level be lowered or that the reservoir be emptied to allow access to deposits by earth-moving equipment. At some sites with predictable seasonal water level variation, dry excavation can be undertaken on a seasonal basis. Disposal area limitations similar to those associated with dredging apply, the difference being that sediment transport by truck haulage will typically be more disruptive than a slurry pipeline. Dry excavation can easily remove commercially valuable coarse material from the delta, but removal of deep deposits of poorly consolidated fine sediment presents significant difficulties absent a period for dewatering and consolidation. For large projects dry excavation is normally more costly than dredging.

Sediment Removal by Flushing

Pressure flushing occurs when a submerged low-level outlet is opened to release sediment while the reservoir level is high, producing a localized scour cone immediately above the pressure flushing outlet. This technique can be used to keep the

Photo 7.3 Dry Excavation at the Pellejas Hydropower Diversion Dam in Puerto Rico

Figure 7.11 Localized Scour Cone Created by Pressure Flushing

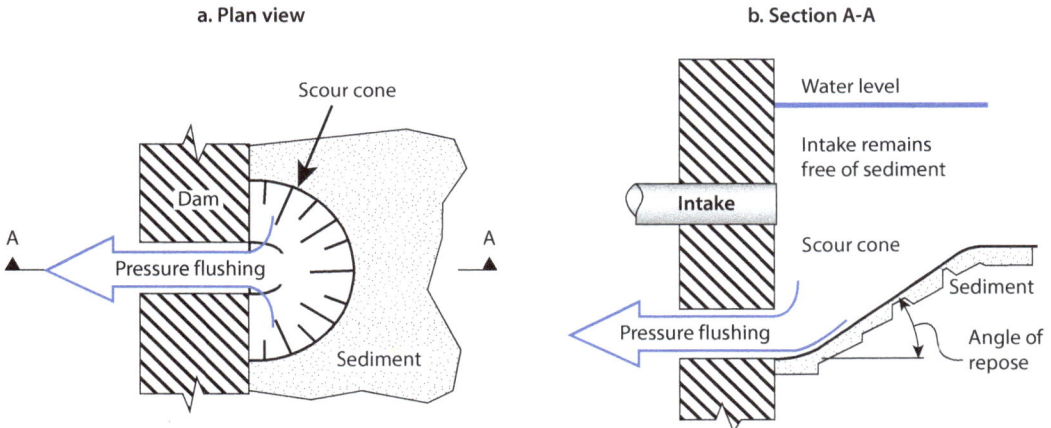

a. Plan view

b. Section A-A

immediate vicinity of an intake free of sediment, as illustrated in figure 7.11. In granular sediment the angle of repose of the scour cone under continuously submerged conditions will approximate the submerged angle of repose of the sediment, on the order of approximately 30 degrees. In the case of cohesive sediment this angle can be different.

Extending the Life of Reservoirs • http://dx.doi.org/10.1596/978-1-4648-0838-8

Empty flushing, or simply *flushing,* entails opening a low-level outlet to completely empty the reservoir, thereby scouring sediment deposits. *Sequential flushing* occurs when two or more reservoirs in series are flushed simultaneously; water is released from an upper reservoir to scour sediment from the lower one, and sediment released from the upper reservoir(s) passes through the downstream reservoirs with minimal redeposition. A more detailed review of flushing is given by White (2001) and Atkinson (1996).

The generalized sequence of a flushing event is schematically illustrated in figure 7.12. As the reservoir is drawn down at the initiation of the flushing event, sediment from upstream is scoured, reworked, and moved progressively closer to the dam as the pool level drops. When the level drops so that high flow velocity is sustained along the entire length of the reservoir, the reworked sediments exit the low-level outlet as a thick muddy flow, creating a high spike in suspended sediment concentration. The concentration drops quickly as the easily eroded sediment is removed from the flushing channel and rate of scour stabilizes. This variation in discharged water quality is conceptually outlined in panel b of figure 7.12. Note that, when a reservoir with consolidated sediments is flushed for the first time, the peak concentrations are typically lower than in a reservoir that is flushed regularly, but the high-concentration flow is sustained for as long as the flushing channel is actively being eroded. For regular flushing, the sediment deposited into the flushing channel each year does not consolidate. It is rapidly mobilized and discharged

Figure 7.12 Flushing Event and Quality of Discharged Water

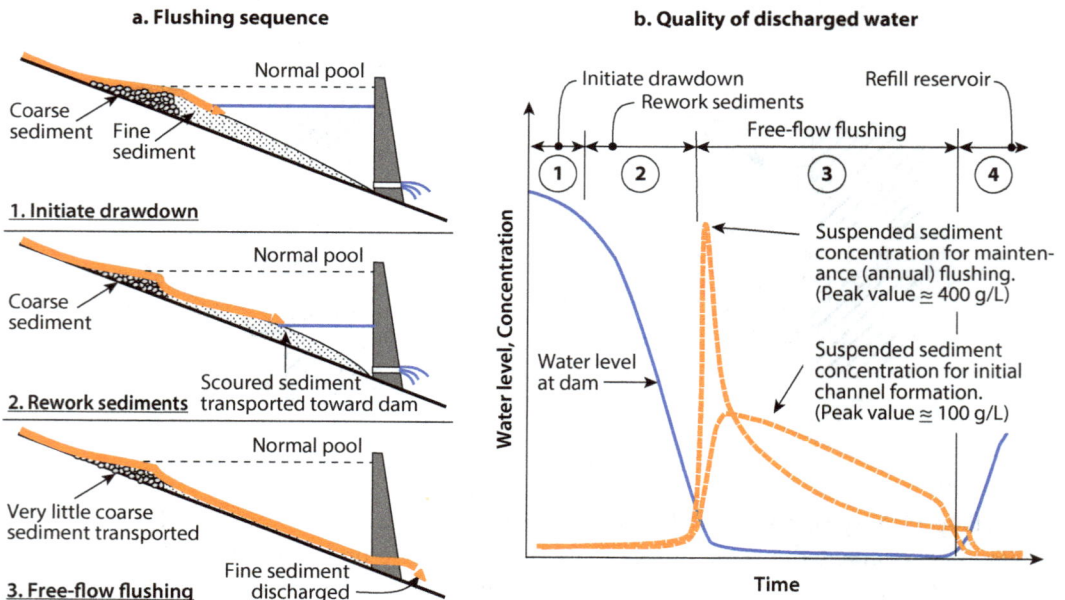

a. Flushing sequence

b. Quality of discharged water

Source: Adapted from Morris and Fan 1998.
Note: g/L = grams per liter.

as soon as free flow exists through the low-level outlet, resulting in an extremely high spike in suspended sediment concentration.

Flushing releases high sediment loads with limited water volumes, frequently producing downstream environmental impacts including low dissolved oxygen, high sediment concentration that interferes with the function of gills and smothers stream benthos, reduction in visibility and light penetration, and channel morphological impacts such as infilling of pools and clogging of river gravels with fine sediment, thereby eliminating spawning sites and habitat. Social and economic impacts include interference with water treatment processes for municipal or other users, sedimentation within irrigation canals if not designed to transport sediment, accumulation in heat exchangers that draw water from the river, reduction of recreational quality, impacts to fisheries of economic importance, accumulation in flood control and navigational channels, and impacts to coastal areas. Although the total amount of sediment released is not different from what would have been transported downstream absent the dam, the combination of high sediment concentrations during flushing, changed downstream hydrology due to the dam, and the potential to release sediment-laden water out of sync with the natural hydrologic and biological cycles can produce large adverse impacts.

The maximum instantaneous suspended sediment concentration in water flushed from a reservoir with fine sediment accumulation may exceed 100,000 milligrams per liter. In contrast, in small reservoirs impounded by large gates (such as a barrage constructed for hydropower ponding), and that have accumulated predominately coarse sediment, the maximum increase in suspended sediment concentration during flushing may be as small as five milligrams per liter when drawdown is controlled and a large dilution flow is available (Espa et al. 2014). Measures to minimize the adverse environmental impacts of reservoir flushing include optimizing the timing of flushing release to avoid environmentally sensitive periods (such as spawning), providing large dilution flows from either natural runoff events or releases from other dams, and flushing more frequently so that each event releases a smaller amount of sediment that can be more readily assimilated by the downstream environment.

The long-term volume that can be sustained by flushing is limited by the width and depth of the flushing channel. Figure 7.13 illustrates the basic geometry of the flushing channel in panel a and maintenance of the flushing channel cross-section while sedimentation continues on adjacent off-channel terraces in panel b. In narrow reservoirs it may be feasible to sustain most of the original volume, but in wide reservoirs only a small portion of the original volume may be sustained free of sediment. Because of the limited duration and discharge of flushing events, the coarse fraction of inflowing sediment that is delivered to the reservoir by large flood events may continue to accumulate. In this case, flushing will not create a complete sediment balance across a reservoir, and it will eventually fill with coarse material despite the control of fine sediment by flushing.

Extending the Life of Reservoirs • http://dx.doi.org/10.1596/978-1-4648-0838-8

Figure 7.13 Cross-Sections of Flushing Channel

a. Basic flushing channel geometry

b. Flushing channel over time

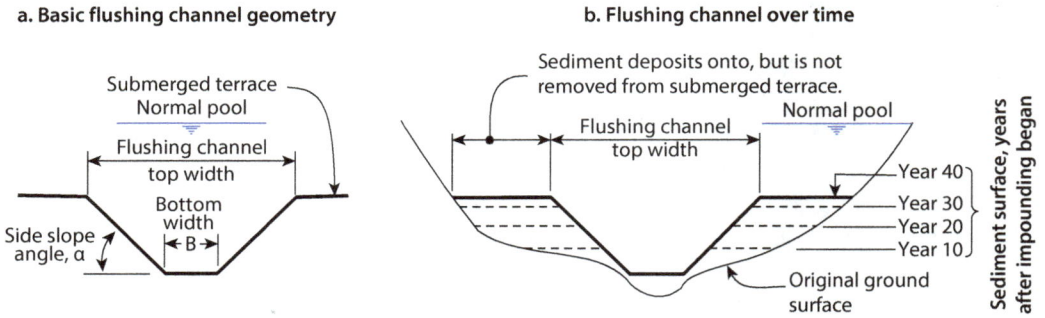

The submerged terrace outside of the channel scoured by flushing will
continue to accumulate sediment, although the rate of accumulation will be
greatly reduced because turbid density currents will travel along and deposit
sediment within the flushing channel. These turbid density current deposits
can be easily removed during the next flushing event. If a low-level outlet or
turbidity siphon is placed at the downstream end of the flushing channel, the
current that flows along the flushing channel may be vented beyond the dam.
Sediment deposition onto the submerged terrace in figure 7.13 can be minimized if the reservoir level is drawn down to prevent terrace submergence
during periods of high sediment inflow.

Management Options and Reservoir Capacity

A variety of management options may be applicable within a region, as outlined for Japan by Sumi and Kantoush (2010). More than one technique may
also be applied at a given reservoir, either sequentially or concurrently. The
applicability of management techniques can depend on the reservoir's hydrologic capacity. A reservoir's hydrologic capacity may be expressed as the
capacity-inflow (C:I) ratio (for example, as used in the Brune curve, figure 5.1),
which is the same as the *retention time* expressed in years for the reservoir at
full capacity. It is the dimensionless ratio of reservoir capacity to mean annual
flow (MAF) entering the reservoir. For example, a C:I ratio of 0.5 means that
the reservoir volume is equivalent to half of the mean annual runoff from its
tributary watershed. Most reservoirs have a capacity less than 0.5 MAF, but
reservoirs in semi-arid regions with highly variable inflows may have a volume
equal to multiple years of inflow. Hydrologic capacity is independent of the
absolute size of the reservoir. For example, although the initial capacity of
Pakistan's large Tarbela reservoir was 14.3 cubic kilometers, this volume is
only 19 percent of the mean annual inflow of the Indus River (C:I ratio =
0.19). Since 1975 Tarbela has lost more than one-third of that capacity to
sedimentation.

A reservoir's hydrologic capacity heavily influences which sediment management techniques are applicable. For example, a hydrologically small reservoir cannot store floods; it will release floodwater over its spillway. This is water that could be used instead to release sediment downstream. Small reservoirs can also be emptied for flushing and rapidly refilled. In contrast, a large reservoir (for example, C:I > 0.5) will infrequently release floodwater, and emptying for flushing will rarely be practical. Figure 7.14 shows the general range of applicability of different management techniques as a function of the reservoir's current hydrologic capacity, which diminishes with time as a result of sedimentation. In figure 7.14 the sedimentation rate is computed on the vertical axis as the reservoir volume divided by annual sediment load (expressed in terms of volume loss). For example, a reservoir life of 100 years corresponds to a 1 percent annual rate of storage loss. The horizontal axis corresponds to the C:I ratio.

Figure 7.14 Applicability of Sediment Management Techniques Based on Hydrologic Capacity and Sediment Loading

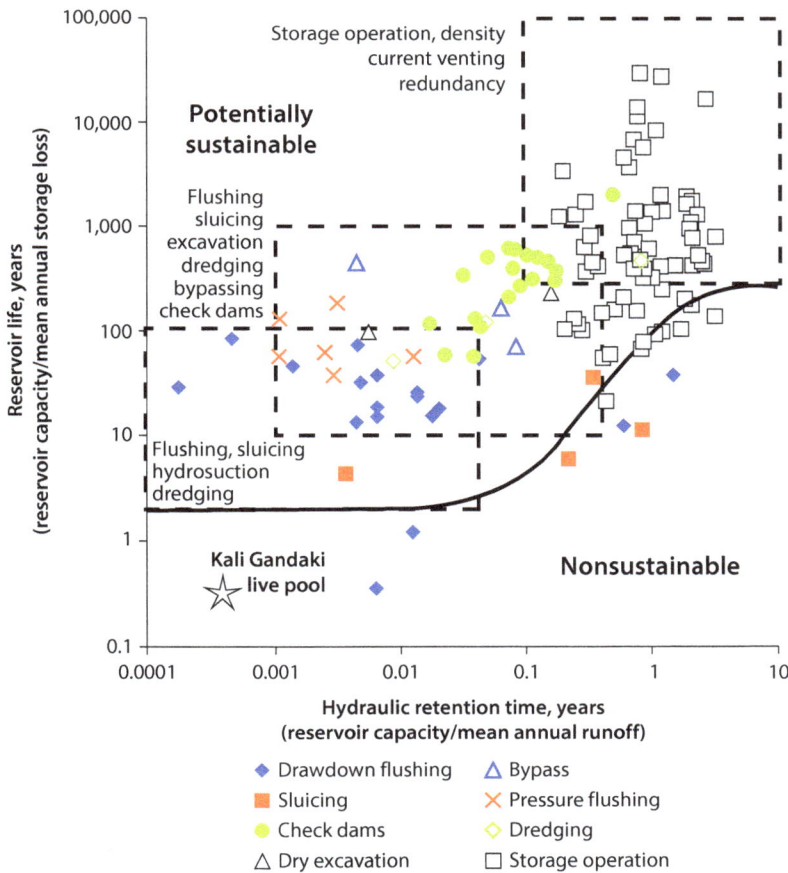

Source: Modified from Annandale 2013.

This figure provides a general guideline for the types of management strategies that may be feasible given a reservoir's current capacity, and which techniques will become more viable as a reservoir's hydrologic capacity is diminished by sedimentation. Conditions at every reservoir are different, and figure 7.14 should only be used as a general guideline; it is not a definitive design tool. For example, the Kali Gandaki hydropower dam in Nepal successfully maintains daily regulation storage volume (pondage) by sluicing, even though it falls into the "nonsustainable" region of figure 7.14.

Adaptive Strategies

Adaptive strategies are actions to mitigate the impacts of sedimentation but that do not involve handling the sediment. They may be used along with or instead of active sediment management. Several types of adaptive strategies are outlined below.

- *Reallocate storage and improve operational efficiency.* Multipurpose reservoirs may be divided into two or more beneficial pools defined based on water level. For example, a reservoir may have a high-level normally empty pool reserved for capturing flood flows, and a lower-level normally full water conservation pool used for water supply storage (figure 7.15). The lowest pool, dead storage, may be allocated to "sediment storage," although sedimentation will normally affect all pools. However, sedimentation does not affect all pools equally, and in many reservoirs the flood control storage pools have experienced much less sedimentation than the lower pool(s) used for water supply, especially in areas where sediment inputs are primarily composed of fines. As a result, sedimentation will affect water supply much more quickly than it will affect flood control. Pool limits may be modified to reallocate the storage loss in a more equitable manner among users so that sedimentation affects both pools to the same degree. This *pool reallocation* can be accomplished by adjusting the boundary limit between the two pools, thereby raising the elevation of the top of the conservation pool at the expense of the flood control pool.

Figure 7.15 Allocation of Flood Control and Conservation Pools in a Multipurpose Reservoir

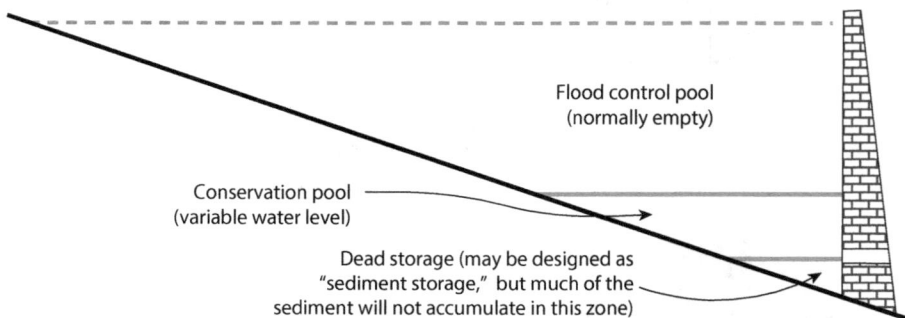

Flood control pool
(normally empty)

Conservation pool
(variable water level)

Dead storage (may be designed as
"sediment storage," but much of the
sediment will not accumulate in this zone)

It may also be possible to improve flood control efficiency, for example, by refining the reservoir operating rules to optimize utilization of the available storage, replacing a decades-old operating rule with a modern operating rule based on real-time hydrologic data. In some regions, the conjunctive use of surface and ground water may be an effective strategy for reducing the impact of storage loss by sedimentation. In hydropower reservoirs, as storage is lost the operating rule can be modified to maximize energy production, progressively raising the minimum operating level and moving the power operation closer to run-of-river operation. Improvements in operational efficiency are typically very economical compared with many types of active sediment management, or the construction of new dams.

- *Modify structures to avoid sediment.* Sediment accumulation will eventually reach critical structures and equipment including spillways, intakes, and hydromechanical equipment. These components may be modified to handle the sediment, for example, by raising or otherwise modifying intakes, by providing protective coatings to hydromechanical equipment, or other measures.
- *Raise dam to increase volume.* Storage may be increased by raising the dam or constructing new storage, thereby temporarily offsetting the storage loss. A new replacement reservoir may be constructed that incorporates a sustainable design.
- *Water loss control and conservation.* Water supply systems frequently contain multiple opportunities to increase water use efficiency, sustaining productivity while using less water. This may include water use conservation, water reuse, and similar techniques. Water-intensive low-value activities may be eliminated. This strategy provides considerable opportunities for addressing water shortages from drought, climate change, and reservoir sedimentation.
- *Decommission infrastructure.* The long-term sustainable use of all reservoirs is not justified, and a dam may be decommissioned when sedimentation renders its continued operation no longer economic or otherwise of sufficient benefit. However, this decommissioning must plan for the long-term management of sediment. For example, will sediment flowing over the dam eventually endanger the structure? Will the delta continue to grow upstream and threaten upstream communities or land uses? Should the dam be modified or removed to restore environmental conditions along the river? What is the fate of the sediment released by dam removal?

Sediment Modeling Approaches

Several different sediment modeling approaches may be applicable for the analysis of sustainable management strategies for new projects, as well as the reconfiguration of existing projects to incorporate sediment management for sustainable use. Both numerical and physical modeling of sediment transport may be performed to refine design parameters to produce a more cost-effective and sustainable project. Modeling is arguably the most important aspect of hydraulic design, and is the best tool available for determining and confirming design parameters.

Extending the Life of Reservoirs • http://dx.doi.org/10.1596/978-1-4648-0838-8

Millions of construction dollars and long-term operational expenses have been eliminated as a result of project optimization through modeling.

Even if a model was prepared before project construction, once the project is in operation it is often useful to revisit the sediment management strategy in light of monitoring data, which may differ from conditions anticipated during design. Significant uncertainty about the rate of sediment delivery into the reservoir may arise during design, particularly concerning the inflow of coarse bed material, and the problems posed by coarse sediment may be different from the original modeling assumptions.

Conceptual Model

The initial and most important modeling step is development of the conceptual model in the minds of the modelers and designers. This conceptual model is essential for identifying the problems that need to be analyzed in more detail, and for selecting the analytical approaches to be used. It is also necessary for developing key modeling parameters such as the sediment grain sizes to be simulated, transport rates of fine and coarse bed material, and so on. It is important that the modelers visit the project site either before starting or during the earliest stages of modeling to improve their understanding of the system they are simulating. This site visit should correspond to the period of low flow so that the sediment beds in the river are exposed and available for inspection and determination of the bed material grain size.

Numerical Modeling

Numerical modeling involves the construction of a computer model to simulate sediment behavior under different operational scenarios based on sediment transport equations. Both one-dimensional (1D) and two-dimensional (2D) numerical modeling may be performed. One-dimensional modeling simulates the reservoir in a linear manner, representing the reservoir as a series of cross-sections. Water and sediment are transported from one cross-section to the next, but no lateral movement from one side of the reservoir to the other can be simulated because conditions are averaged across the entire cross-section. While this represents an obvious simplification to the real system, because sediments tend to be carried downstream with the predominant current along the reservoir, setting up a 1D model that simulates the flow along the main flow path within the reservoir can give a good approximation of sediment behavior in many reservoirs.

One-dimensional models are commonly used for the evaluation of long-term (for example, 100-year) sedimentation patterns, and to examine how design and operational alternatives influence long-term evolution of the reservoir. Because 1D models are also commonly used to analyze river sedimentation problems, they normally include features such as the computation of multiple grain sizes simultaneously and simulation of an *armor layer* of coarse sediment on top of sediment deposits, which can impede the transport of bed material during periods of normal flow. This modeling capability may become an important consideration when simulating behavior of the reservoir delta and the long-term equilibrium condition with respect to sediment deposition and scour. One limitation is that 1D models normally do not simulate the progress of turbid density currents through a reservoir, and thus may underestimate the transport of fine sediment into the area of the dam.

Two-dimensional modeling uses a grid to simulate the reservoir geometry and can simulate the lateral movement of water and sediment across the cross-section, as well as the downstream movement. Two-dimensional models are much more computationally intensive than 1D models, resulting in much longer computer run times. Therefore, 2D models have historically been limited to simulating areas of the reservoir where the lateral movement of water and sediment is of particular importance, such as the area near intakes and spillway and evolution of a delta. They may also be used for short simulation periods, such as individual flood events, as opposed to decades-long time series. However, increasing computer capacity is expanding the range of uses for multidimensional models.

Physical Models

Physical modeling involves preparation of a scale model of the prototype project. Sediment is simulated in the model using either natural sediment of smaller diameter, or a less dense material including plastic beads and ground walnut shells. A number of parameters are important for determining the rate and pattern of sediment transport, for example, water depth, velocity, shear stress of water against the bed, and sediment diameter. These parameters do not all scale at the same rate, and the physical modeling scale includes trade-offs for best simulating project behavior.

Physical models are commonly used to examine the three-dimensional (3D) details of flow and sediment transport conditions at intakes and spillways, including development of scour downstream of the dam by the spillway discharge. These questions involve complex secondary flow patterns and turbulence, which are not easily addressed by current numerical models. An example of a physical model of a dam and intake structure is given in photo 7.4. The physical model allows the designer to directly observe flow and sediment transport patterns, and to rapidly make geometric changes in the configuration of project structures and immediately observe their impact on sediment transport patterns. It is an unparalleled tool for allowing both design and management personnel to visualize the system. As a limitation, once the modeling is completed and the model destroyed, it may cost several hundred thousand dollars to reconstruct the model to address any new questions that arise. Careful design and execution of physical modeling is crucial so that important questions are not left unanswered. Contracts for physical modeling must specify how long the model should be maintained before it is dismantled.

Although 3D numerical modeling can also simulate complex flow conditions, current 3D numerical modeling tools are not comparable to physical models in their ability to allow project designers to observe flow and transport patterns, or the ease with which alternative structures can be visualized and tested during model operation. Also, physical modeling is a time-proven technique that

Photo 7.4 Physical Model of Kali Gandaki Dam, Intake and Sedimentation Basin

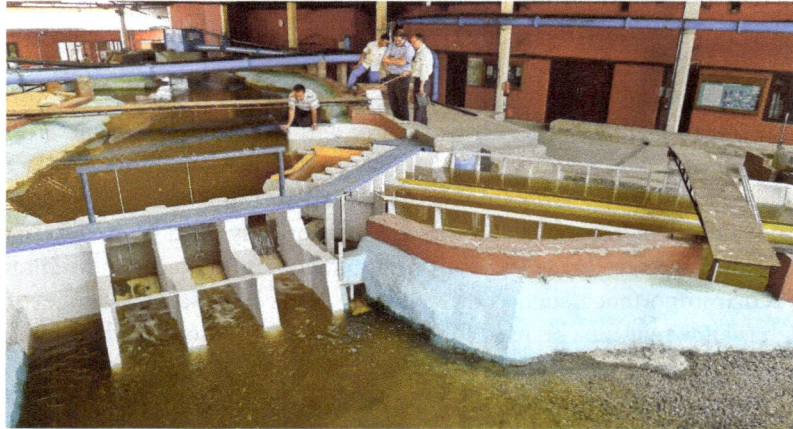

Source: © G. Morris. Used with permission. Further permission required for reuse.
Note: Model by Hydrolab, Kathmandu, for Nepal Electric.

designers are familiar with. For the foreseeable future physical modeling is expected to remain the tool of choice for simulating complex flow fields around dam and intake structures and for analyzing spillway discharge.

Sequence of Modeling Studies

The normal sequence in modeling is to start from the most general model and proceed with increasing detail and complexity as the design is refined. Following this sequence, modeling normally starts with a 1D numerical simulation to analyze the overall rate and pattern of sedimentation, the grain size that will be transported to the area of the intake or outlet works, and how these factors will change over time. This step is sometimes followed by 2D numerical modeling and physical modeling. If both 2D and physical modeling are to be performed, the logical sequence is to perform the 2D numerical modeling first, to be used as the basis for determining the design (or design alternatives) to be investigated by physical modeling. To gain time, the 1D and 2D models can be developed simultaneously, but the 1D numerical modeling results will normally be used to establish the sediment transport input for the 2D model of the area closer to the dam.

Physical modeling is normally performed for dams having significant discharges to aid in spillway design and to optimize the configuration and placement of the intake and sediment sluicing features such as low-level outlets or deep crest gates. The release of floating debris is another design consideration evaluated by physical modeling. Although 2D numerical modeling may be considered optional, physical modeling is normally considered to be a requirement for finalizing the design of projects of moderate or greater complexity or size.

References

Allen, Peter M., and John A. Dunbar. 2005. "Dredging vs. New Reservoirs." Texas Water Development Board, Austin, TX.

Annandale, G. W. 2013. *Quenching the Thirst: Sustainable Water Supply and Climate Change.* Charleston, SC: CreateSpace.

Atkinson, E. 1996. *The Feasibility of Flushing Sediment from Reservoirs.* Report OD-137. Wallingford: HR Wallingford.

Auel, C., T. Berchtold, and R. Boes. 2010. "Sediment Management in the Solis Reservoir Using a Bypass Tunnel." Proceedings of the 8th ICOLD European Club Symposium, Innsbruck, September 22–23.

Auel, C., R. Boes, T. Ziegler, and C. Oertli. 2011. "Design and Construction of the Sediment Bypass Tunnel at Solis." *Hydropower and Dams* 3: 62–66.

Boes, R. W. 2015. "Proceedings of the First International Workshop on Sediment Bypass Tunnels." Laboratory of Hydraulics, Hydrology and Glaciology, ETH, Zurich, Switzerland.

Chanson, Hubert. 2004. "Sabo Check Dams: Mountain Protection Systems in Japan." *International Journal of River Basin Management* 2 (4): 301–7.

Desta, L., and B. Adugna. 2012. *A Field Guide on Gully Prevention and Control.* Nile Basin Initiative. http://www.bebuffered.com/downloads/ManualonGullyTreatment_TOTFinal_ENTRO_TBIWRDP.pdf.

De Vente, J., J. Poesen, M. Arabkhedri, and G. Verstraeten. 2007. "The Sediment Delivery Problem Revisited." *Progress in Physical Geography* 31 (2): 155–78. doi:10.1177/0309133307076485.

Espa, P., G. Crosa, G. Gentili, S. Quadroni, and G. Petts. 2014. "Downstream Ecological Impacts of Controlled Sediment Flushing in an Alpine Valley River: A Case Study." *River Research and Applications* 31 (8): 931–42.

Ffolliott, Peter F., Kenneth N. Brooks, Daniel G. Neary, Roberto Pizarro Tapia, and Pablo Garcia Chevesich. 2013. *Soil Erosion and Sediment Production on Watershed Landscapes. Processes, Prevention, and Control.* Montevideo, Uruguay: UNESCO.

Gellis, Allen C., Andres Cheama, Vanissa Laahty, and Sheldon Lalio. 1995. "Assessment of Gully-Control Structures in the Rio Nutria Watershed, Zuni Reservation, New Mexico." *Journal of the American Water Resources Association* 31 (4): 633–46.

Geyik, M. P. 1986. *FAO Watershed Management Field Manual: Gully Control.* Rome: FAO. http://www.fao.org/docrep/006/AD082E/AD082e00.htm.

Heede, B. H. 1966. "Design, Construction and Cost of Rock Check Dams." U.S. Forest Service Research Paper RM-20, Fort Collins, CO.

———. 1978. "Designing Gully Control Systems for Eroding Watersheds." *Environmental Management* 2 (6): 509–22.

———. 1982. "Gully Control: Determining Treatment Priorities for Gullies in a Network." *Environmental Management* 6 (5): 441–51.

Hotchkiss, R., and X. Huang. 1995. "Hydrosuction Sediment-Removal Systems (HSRS): Principles and Field Test." *Journal of Hydraulic Engineering* 121 (6): 479–89.

Kantoush, S. A., and T. Sumi. 2010. "River Morphology and Sediment Management Strategies for Sustainable Reservoir in Japan and European Alps." No. 53B, Annuals of Disaster Prevention Research Institute, Kyoto University, Japan.

Kung, Chen-Shan, Min-Yi Tsai, Yi-Liang Chen, Shih-Wei Huang, and Ming-Yang Liao. 2015. "Sediment Sluicing Tunnel at Nanhua Reservoir in Taiwan." In Proc. *First International Workshop on Sediment Bypass Tunnels*, VAW-Mitteilungen 232. edited by R. M. Boes, 71–83. Laboratory of Hydraulics, Hydrology and Glaciology, ETH Zurich, Switzerland.

Leopold, L. M., M. G. Wolman, and J. P. Miller. 1964. *Fluvial Processes in Geomorphology*. New York: W. H. Freeman.

Marui, H. 1988. *FAO Watershed Management Field Manual: Landslide Prevention Measures*. Rome: FAO. http://www.fao.org/docrep/006/S8390E/s8390e00.htm.

Morris, G. L. 2010. "Offstream Reservoirs for Sustainable Water Supply in Puerto Rico." American Water Resource Association, Summer Specialty Conference, San Juan, August 30–September 1. American Water Resources Assn., Middleburg, Virginia.

———. 2015. "Management Alternatives to Combat Reservoir Sedimentation." In Proceedings of the International Workshop on Sediment Bypass Tunnels, Zurich, April 27–29.

———, and J. Fan. 1998. *Reservoir Sedimentation Handbook: Design and Management of Dams, Reservoirs and Watersheds for Sustainable Use*. New York: McGraw-Hill.

Renwick, W. H., S. V. Smith, J. D. Bartley, and R. W. Buddemeier. 2005. "The Role of Impoundments in the Sediment Budget of the Conterminous United States." *Geomorphology* 71 (1–2): 99–111. doi:10.1016/j.geomorph.2004.01.010.

Rosgen, D. L. 1994. "A Classification of Natural Rivers." *CATENA* 22: 169–99.

———. 1996. *Applied River Morphology*. Pagosa Springs, CO: Wildland Hydrology Books.

Sumi, Tetsuya, and Sameh Ahmed Kantoush. 2010. "Integrated Management of Reservoir Sediment Routing by Flushing, Replenishing, and Bypassing Sediments in Japanese River Basins." International Symposium on Ecohydrology, Kyoto, 831–38.

Sumi, Tetsuya, Masahisa Okano, and Yasufumi Takata. 2004. "Reservoir Sedimentation Management with Bypass Tunnels in Japan." Proceedings of the 9th International Symposium on River Sedimentation, Yichang, China, October 18–21.

Sumi, T., T. Yoshimura, K. Asazaki, and T. Sato. 2015. "Retrofitting and Change in Operations of Cascade Dams to Facilitate Sediment Sluicing in the Mimikawa River Basin." Q99–R45, 597–611. 25th Congress, International Commission on Large Dams, Stavanger, Norway.

Turner, T. M. 1996. *Fundamentals of Hydraulic Dredging*. New York: ASCE Press.

U.S. Federal Interagency Stream Restoration Working Group. 1998. "Stream Corridor Restoration: Principles, Processes, and Practices." Federal Interagency Stream Restoration Working Group (FISRWG) (15 Federal agencies of the U. S. government). GPO Item No. 0120-A; SuDocs No. A 57.6/2:EN 3/PT.653. ISBN-0-934213-59-3.

Valentin, C., J. Poesen, and Y. Li. 2005. "Gully Erosion: Impacts, Factors and Control." *CATENA* 63 (2–3): 132–53.

Walling, Des E. 1983. "The Sediment Delivery Problem." *Journal of Hydrology* 65: 209–37.

White, Rodney. 2001. *Evacuation of Sediments from Reservoirs*. London: Thomas Telford Ltd.

CHAPTER 8

Sediment Management at Run-of-River Headworks

Gregory L. Morris

Introduction

A run-of-river (ROR) hydropower plant generates power from the daily flow of a river, but may include operational storage limited to the *pondage volume* needed for daily flow regulation. During the dry season this pondage allows the plant to accumulate water during off-peak hours to operate at full power during peak demand, typically on the order of about six hours a day. Because of their small storage volume (if any), ROR projects are challenged by high sediment loads starting very early in their operational life. Many plants suffer elevated operational costs and accelerated rates of turbine abrasion as a result. ROR plants may operate under heads from a few meters to over 1,000 meters, and the sensitivity of the turbine runners to abrasion by sediment increases as a function of increased hydraulic head (Nozaki 1990).

This chapter outlines basic concepts to consider in the design or rehabilitation of ROR headworks to improve their function, including (1) basic headworks components, (2) fluvial morphology and site selection, (3) types of intakes, (4) flow control and conveyance, (5) sediment removal from diverted water, and (6) sediment-guided operation and monitoring.

As existing storage reservoirs fill with sediment, they may transition from storage to ROR operation. This change will normally require reconfiguration of the dam, intake, and operating rule to allow power production to continue even after the delta with its load of coarse sediment has reached the dam. The concepts presented in this chapter are also relevant to tall dams designed from the onset to shift from impounding to ROR after a period of decades when the delta reaches the intake, as well as to the redesign of storage reservoirs that will be converted to sustainable ROR operation when their storage volume is lost to sedimentation.

Configurations of ROR Hydropower Plants and Objectives of Headworks Design

The basic features of ROR headworks are illustrated in figure 8.1, which includes a weir or dam in the river to provide additional head and sufficient water depth in front of the intake, sluice gates for scouring accumulated sediment from the intake area, an intake structure that admits flow while minimizing the capture of sediment and debris (maximizing the continued downstream transport of both sediment and debris), a trash rack to remove debris from the diverted water, and sedimentation basins to remove sand from the water used for power.

Performance standards for ROR headworks may be organized into the five categories shown in figure 8.2. To minimize operational costs intakes should be designed to (1) safely pass large or extreme floods, (2) pass waterborne debris and ice, (3) pass sediment and minimize damage from extreme events such as debris flows, (4) avoid the buildup of bed material in front of the intake, and (5) minimize the entrainment into the intake of suspended sediment and, where pertinent, air. The headworks design should facilitate placement of stoplogs in the intake for maintenance purposes and allow access by equipment to all parts of the structure during the low-flow season. Environmental considerations such as flow maintenance and fish passage are also essential design and operational considerations at many intakes.

ROR intakes typically have three streamflow-dependent operational regimes, as outlined in table 8.1. At lower flows the sediment concentration is limited and generally does not pose significant problems, except in sand-bed rivers since sand is always in motion. Both sediment concentration and suspended grain size

Figure 8.1 Principal Components of Run-of-River Headworks Relevant to Sediment Management

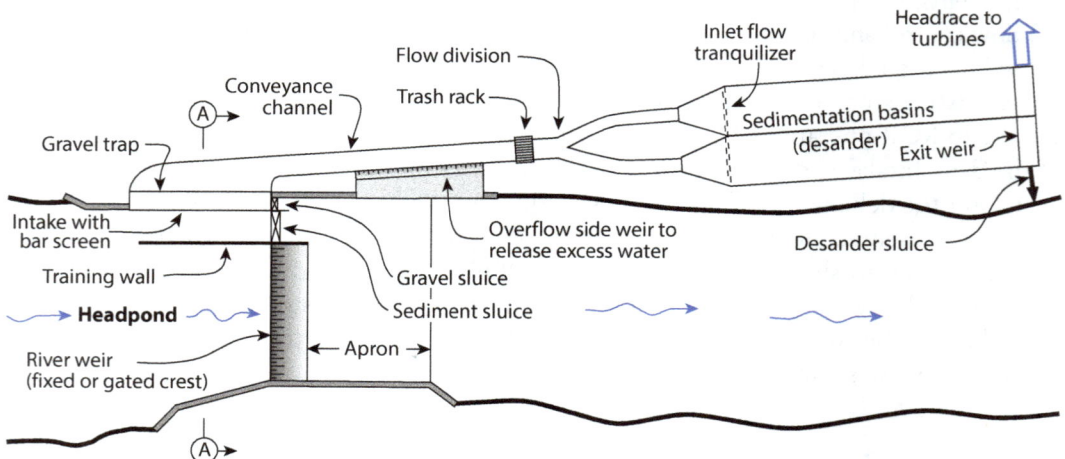

Figure 8.2 Performance Standards for Run-of-River Headworks

Performance standard	Consequences of noncompliance
1. Passage of floods including hazard floods	High flood hazard
2. Passage of ice, trash, and floating debris	Unsafe and damaging conditions during normal operations
3. Passage of sediments	Unsafe and damaging conditions during normal operations
4. Bed control at intake	Unsafe and damaging conditions during normal operations
5. Exclusion of suspended sediment and air	Unsafe and damaging conditions during normal operations

Less consistent operation and higher maintenance costs
compared with similar plants meeting perfromance standards

Source: Based on Haakon Støle 2014 (personal communication).

Table 8.1 Operational Ranges Characteristic of Run-of-River Power Plants

Streamflow	Operation
< Design + environmental flow (very low sediment and debris load)	All water diverted to power + environmental flow. Sluice sediment from in front of the intake only as necessary.
> Design + environmental flow	Continuously discharge over a fixed weir, open gates to sluice excess water and sediment.
> Maximum operational flow (very high sediment and debris load)	Intake out of service due to high sediment and debris load.

captured by the intake tend to increase with discharge, and both designers and operators should pay particular attention to intake performance at high flows. An important design objective is to optimize the intake configuration to sustain full power production at the highest discharge possible.

Diverted water is normally passed first through a gravel trap at the intake before conveyance into a sedimentation basin to remove larger sand particles, typically exceeding 0.15 or 0.20 millimeters in diameter, before being conducted to the power house along the headrace. The trapped sediment is flushed from the sedimentation basin either continuously or by periodic emptying and scouring. For intakes feeding directly to a power tunnel, the exclusion of air is an important design consideration and the intake may require submergence to prevent vortex formation and air entrainment. This is a typical arrangement in a storage reservoir, but is not generally an appropriate design for an ROR intake. A preferred alternative is a high-level intake discharging into an open channel to eliminate the problem of air entrainment, thereby allowing the ROR intake to be placed at the highest level possible to minimize entrainment of fluvial sediment.

Extending the Life of Reservoirs • http://dx.doi.org/10.1596/978-1-4648-0838-8

Fluvial Morphology and Site Selection

Importance of Fluvial Morphology and Sediment Behavior

Hydropower plants can have very long lives, and a number of plants have now been in operation for more than 100 years. To successfully operate engineered structures over long periods in the fluvial environment, the design must incorporate an understanding of stream behavior and sediment transport, including patterns of erosion and deposition that can be anticipated and managed over many decades and during extreme floods.

When the diverted flow consists of a small fraction of the annual high flow, simply locating the intake in a zone of natural scour at a curve in the river may be adequate without constructing any additional sediment handling features. However, when the diverted flow exceeds about 40 percent of the mean discharge, or in sand-bed rivers, active sediment management will become a major consideration for intake design and will normally include the requirement for gated structures that can periodically sluice sediment away from the intake (ASCE 1995).

Fluvial Similarity

An important step in the design process is to visit and understand the strengths and weaknesses of existing intake structures on rivers similar to the design site. However, because there are a variety of river forms, a strategy that works on one river will not necessarily work on another river having different geomorphic characteristics or sediment load. For instance, gravel-bed and sand-bed rivers act quite differently. It is essential that intake sites used for design reference have geomorphic characteristics similar to the design. Factors to consider include bed material, river slope, sand load, geomorphic configuration of the river, and operating head. The Rosgen classification system (Rosgen 1994, 1996) is useful to help evaluate fluvial similarity. Guidelines for applying this classification system may be easily found through an Internet search.

Intake Location

Although rivers are normally thought of as flowing downstream, secondary or rotational currents are also present that move water and sediment from one side of the channel to the other. These secondary currents modify the channel geometry, creating zones of scour as well as zones of sediment deposition that create sand and gravel bars. Secondary currents also contribute to the variation in sand concentration within the wetted cross-section. For example, sand concentration is usually higher near the river bed than at the water surface.

Because of secondary currents, from the standpoint of sediment management locating an intake on the exterior of a river bend is generally preferable. When the current flows against the outer bank of a river bend, the surface water with lower sediment concentration plunges, creating a rotational current that scours a deep pool at the toe of the outer bank as shown in figure 8.3. The flow then crosses the river bed, carrying bed material that is deposited to create the point bar that occupies the interior of the bend. Because of this rotational flow pattern, the concentration of sand entering an intake located at the exterior of a bend is reduced because it abstracts flow from the top of the water column, and the scouring action keeps the intake free of bed material. Conversely, the interior of the bend is an unfavorable intake location because it will have higher sand concentration plus the tendency to accumulate bed material (see figure 8.4).

Intake siting must consider a range of factors including river morphology, geology, access, elevation, geologic hazards posed by adjacent slopes, property ownership, and social and environmental considerations. Headworks design requires compromises, and locating the intake in the preferred location from the standpoint of river morphology may not be possible. Nevertheless, considerable

Figure 8.3 Plunging Flow at Exterior of River Meander

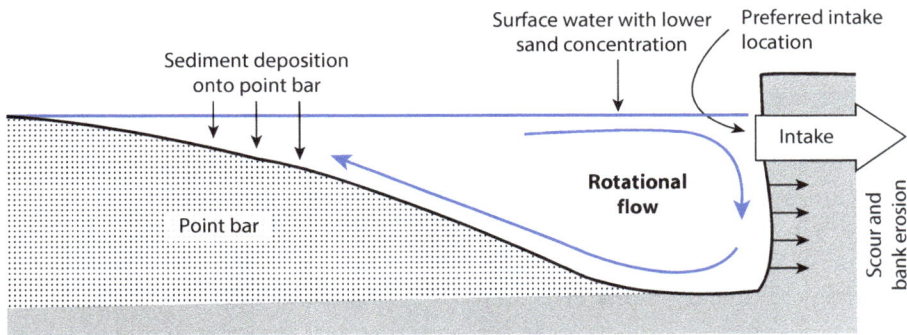

Note: When this rotational flow is superimposed on the downstream flow of the river, the resulting flow path is described as helicoidal. This diagram corresponds to section A-A in figure 8.4.

Figure 8.4 Idealized Schematic of River Meanders and Suitability for Intake Location

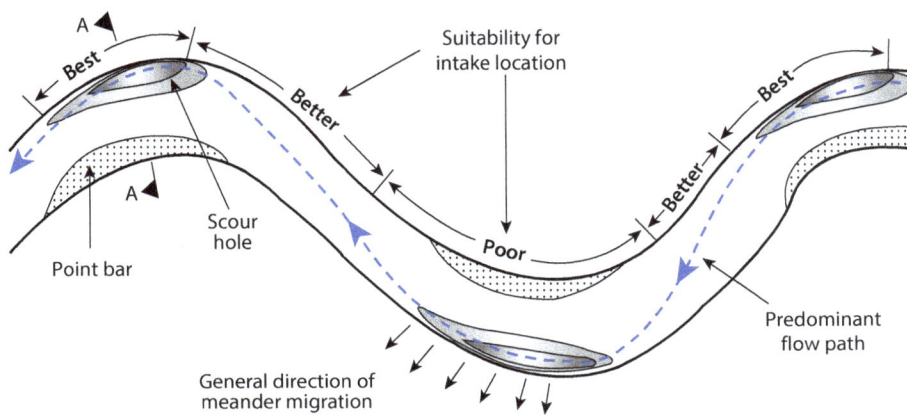

Note: Figure shows general patterns of river meander and suitability for intake location to reduce sediment management problems. The tendency of the river to migrate laterally must also be considered.

Extending the Life of Reservoirs • http://dx.doi.org/10.1596/978-1-4648-0838-8

effort should be made to avoid placement of the intake in a zone having poor geomorphic characteristics that will substantially increase sediment management problems.

Modifying Flow in Front of an Intake

A favorable geomorphic location can minimize the entrainment of sand and produce a scour pattern that keeps the intake face free of bed material. But when the river geometry does not provide favorable hydraulic conditions in front of the intake, the pattern of secondary currents may be improved by optimizing intake orientation, by operating gates, or by constructing river training structures. Orientation of the intake slightly into the flow is essential to help reduce coarse sediment concentration by establishing an impinging flow pattern similar to that illustrated in figure 8.3. An intake that is oriented away from the flow will create an eddy current that can lift sediment from the streambed and into the intake. At larger intakes or intakes with high sediment loads, physical modeling is normally recommended to optimize this aspect of the design. For gated weirs, the flow pattern in the vicinity of the intake may be modified based on the sequence for opening gates as streamflow increases. For example, as river discharge increases, flow curvature and rotational flow may be induced by first opening a bottom-opening spillway gate that is not adjacent to the intake. For new designs, the sequence of gate openings and its impact on sediment entrainment into the intake may be examined by physical model testing. At existing intakes, data from observation and sampling can help operators determine the optimal gate operation, or confirm the results of preconstruction physical modeling. Even under optimal design there will typically be some flow rate (or other criteria such as suspended sediment concentration) beyond which the ROR plant is shut down to minimize abrasion damage. These high flows may be used to flush accumulated sediment from the headpond.

Defensive Design for Extreme Events

ROR hydropower plants are frequently located in mountainous areas where the high stream slope makes it feasible to develop significant hydraulic head over a relatively short distance. These areas can be subject to geologic hazards including landslides, debris flows, and glacial lake outburst floods. Numerous ROR hydropower plants have been damaged by these infrequent but devastating events.

A debris flow is a mass of sediment and water that flows as a thick fluid, transporting mud, soil, and even boulders.[1] In mountainous areas, careful attention should be given to determining the possible impact of debris flows on both the headworks and the power house. Evidence of debris flow hazard may be obtained from historical records and local interviews, experience at similar nearby locations, aerial photography showing landslide scars, and by a geomorphic analysis of the river and its tributaries. Debris flows can transport boulders much larger than those transported by clear-water flows, and the presence

Photo 8.1 Boulder-Strewn Watercourse Resulting from a Debris Flow Event at the Intake to the 30 MW Jagran Power Station in Pakistan-Administered Kashmir

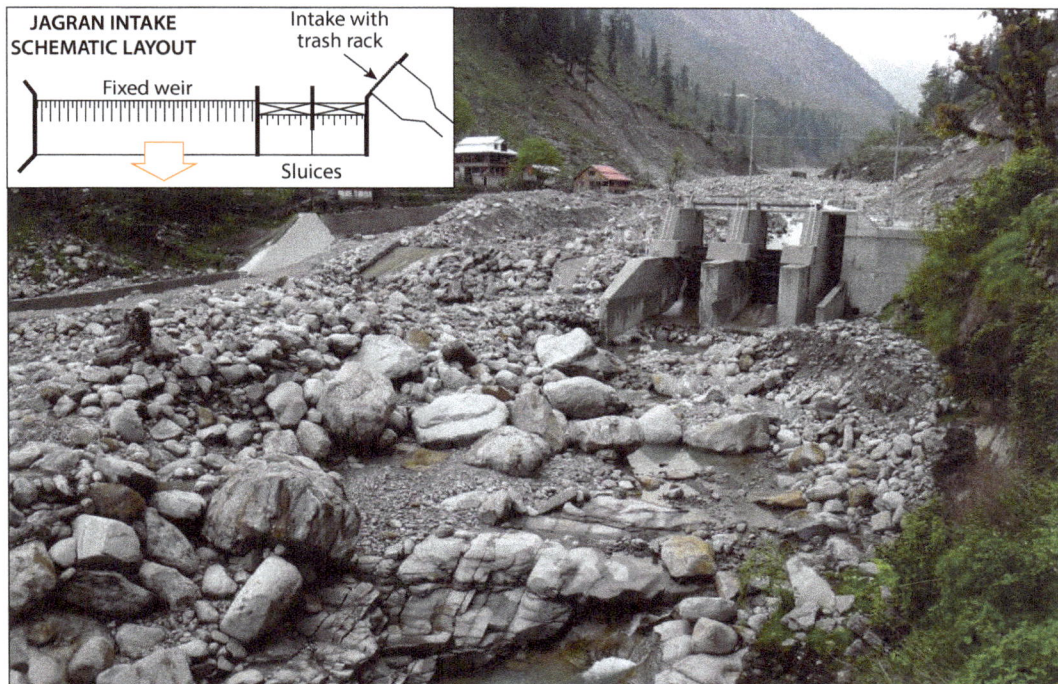

of such material along the river is one indicator of this type of hazard. Photo 8.1 shows an example of a Himalayan intake affected by debris flow with an insert showing the general layout of the intake. Notice that the fixed weir allowed the flow to overtop the structure with relatively little damage to the intake. Had a gated structure been used the damage could have been far worse. This flow carried away a riverside mosque that had reportedly stood for 120 years.

Types of Intakes

Intake configurations may be broadly classified as lateral intakes, bottom intakes, and frontal intakes.

Lateral Intakes

A lateral intake, by far the most common layout, consists of an opening on the side of the river to admit diverted water. It is commonly accompanied by a low dam, either gated or ungated, to elevate the water level, and a sluice gate on the dam adjacent to the intake that can be periodically opened to flush sediment and clear the front of the intake. Lateral intakes may include

Extending the Life of Reservoirs • http://dx.doi.org/10.1596/978-1-4648-0838-8

a bar screen to exclude large debris followed by a trash rack. A gravel trap may be installed immediately after the intake to remove gravel from the diverted flow. This general arrangement is illustrated in figure 8.1. Some intakes also include an undersluice beneath the intake to continuously carry away sediment transported near the bed to prevent it from accumulating and entering the intake.

Frontal Intake

The overall concept behind a frontal intake is to divide the flow into two levels. Water for power production is withdrawn from the top of the flow, while an undersluice continuously releases the bed material and water having higher sediment concentration. This general arrangement is illustrated in figure 8.5. When the reservoir headpond is used to trap sediment, in lieu of a sedimentation basin, a frontal intake can also be used to promote more uniform and parallel flow paths through the sedimentation zone in the headpond.

Bottom Intake

The traditional form of a bottom intake (or drop intake) is the Tyrolean weir, originally used by peasants to divert irrigation water in the Tyrolean Alps, but also used for small hydropower plants. It consists of a submerged weir and inclined bar screen set across the bottom of the stream and used to divert water from mountain torrents. Larger stones and debris are carried over the intake, but sand and gravel (smaller than the screen size) will enter the intake. During low flow periods the intake can abstract all of the water, but flood flows will exceed the intake capacity, as illustrated in figure 8.6. A more recent version of the bottom intake is represented by Coanda-effect screens. To sustain instream environmental flow the required flow may be abstracted upstream of the weir by providing a lateral channel or may discharge across a blind section of the weir, which may be set at a lower elevation to guarantee downstream environmental release before diversion to power.

Figure 8.5 Conceptual Schematic of Frontal Intake Configuration

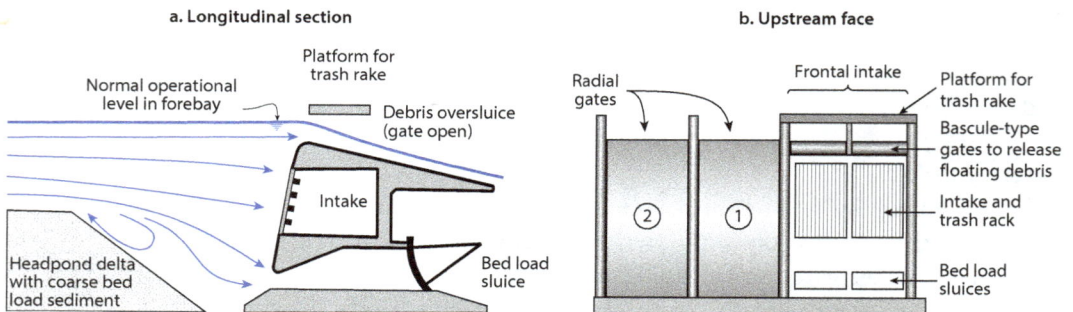

Source: Concept by H. Støle 2014.

Figure 8.6 Conceptual Configuration of Bottom Intake

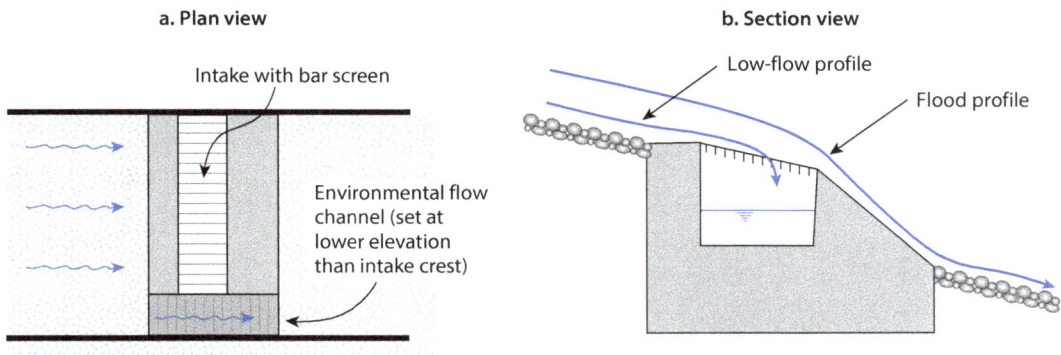

Sediment Management at Headworks

The Value of Storage in ROR Hydropower Plants

ROR plants operate at design capacity in high-flow periods, but in dry periods they may store water during off-peak hours and operate at design capacity during hours of peak power demand, to take advantage of the potentially significant price differential between peak and off-peak power. This limited storage volume used for daily peaking is termed *pondage*. By operating for peak hour production throughout the dry season, the peaking storage may potentially be emptied and refilled through perhaps 100 cycles each year (depending on local hydrology). For example, consider a system in which there is a price differential of $0.01 per kilowatt-hour (kWh) between peak and off-peak hours, and storage is used to accumulate flow for delivery at design discharge during 100 peaking cycles per year on low-flow days. The calculations summarized in table 8.2 show that storage for power peaking can be very valuable, and will typically warrant much greater expenditure for its preservation than the volume of a hydropower storage reservoir that is emptied and refilled only once a year. Peaking storage for high-head plants is particularly valuable. The need for daily power peaking storage capacity is increasing with the growing capacity of intermittent renewable power sources (photovoltaic and wind).

Strategies to Sustain Onstream Peaking Storage

The peaking storage volume in an instream reservoir will fill with coarse bed material unless it is maintained using the methods described in chapter 6, which may include flushing, sluicing, a sediment bypass tunnel, and mechanical excavation. These methods have been used to sustain storage at existing hydropower plants. Given the high value of peaking storage and the smaller hydrologic volume that needs to be maintained, the maintenance of peaking storage is easier and more economically feasible than maintaining the larger volumes required for seasonal storage.

In a narrow reservoir it may be feasible to sustain pondage volume indefinitely by sluicing or flushing. However, in wider reservoirs the flushing channel may occupy only a fraction of the full reservoir width needed to provide the desired pondage volume. In such cases it may be possible to minimize sediment deposition in the zone outside of the flushing channel (recall figure 7.13) by lowering the reservoir level during periods of high sediment inflow. The operational rule would be tailored to the site-specific hydrology. Although this technique reduces the head available for power during the drawdown period, it can sustain pondage volume that may otherwise not be feasible.

Offstream Storage for Power Peaking

Storage for daily regulation may be provided by an offstream storage reservoir having one of the arrangements illustrated in figure 8.7. If the diverted flow is passed through a desander and then into the offstream pondage pool, additional sedimentation will occur in the pondage pool, reducing wear on the turbines but increasing the frequency of sediment removal. If the offstream pondage pool is large, a bypass may not be provided and sediment may be removed by dredging without interfering with hydropower operations.

Alternatively, offstream storage may be situated parallel to the sedimentation basin and used only during low flows (when sediment concentration is low).

Table 8.2 Value of Storage for Daily Peaking Power as a Function of Power Head

	Power head (meters)	
Parameter	100	500
Peaking cycles per year (6 hours duration)	100	100
Price differential, peak vs. off-peak, $/kWh	$0.01	$0.01
Peaking storage required per 1 m³/s of design discharge, m³	16,200	16,200
Annual income increment per 1 m³ of peaking storage	$0.33	$1.63

Note: kWh = kilowatt-hour; m³= cubic meters; m³/s = cubic meters per second. During peaking operation one-quarter of the discharge is from inflow and three-quarters is water released from storage.

Figure 8.7 Arrangement of Offstream Pondage

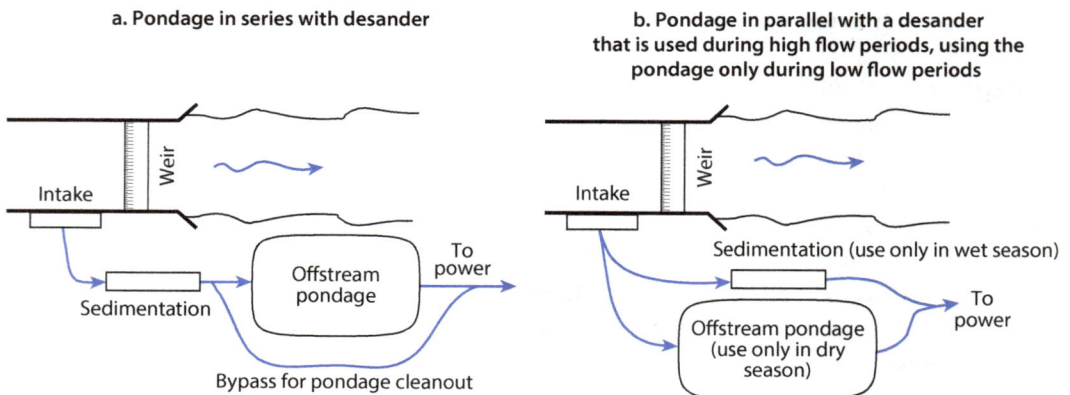

a. Pondage in series with desander

b. Pondage in parallel with a desander that is used during high flow periods, using the pondage only during low flow periods

This configuration provides two alternative flow paths and reduces the sediment load on the pondage volume, but loses the advantage of additional sediment removal.

Abrasion of Headworks by Released Bed Load

Headworks will necessarily pass sediment bed load downstream, and several components are particularly susceptible to abrasion damage. Abrasion-prone components include the sill and invert of outlets used to pass sediment and their corresponding gate structures, the upstream end of piers and guide walls and the lowest one to two meters of these walls adjacent to the floor, and undersluices. Any bends in undersluice channels are zones of particularly high abrasion potential. Because of abrasion and the potential for stones and sediment to lodge in the guide slots of vertical lift gates, radial gates (which do not require guide slots) are better suited for outlets that pass bed material. Examples of damage caused by bed material are illustrated in photo 8.2.

Measures to reduce the rate of wear include (1) provision of at least a 0.5-meter thick high-strength sacrificial concrete without steel reinforcing, (2) use of hard stone granite with joints staggered as shown in panel b of photo 8.2, and (3) use of steel lining. The use of high-strength concrete with annual repair, and steel plate on the lowest meter of sidewalls, has been found to be an economical approach at some sites. The designer needs to ensure that guard gates or stoplogs can be placed upstream, and also downstream if necessary, to enable the abrasion-prone area to be dewatered and repaired during the dry season.

Photo 8.2 Abrasion Damage by Bed Load

| a. Damage to sediment sluice after only six months of service on river transporting gravel and cobbles | b. Damage to sediment sluice, requiring periodic repair of steel gate seat and replacement of granite blocks |

Eroded radial gate seat

Source: G. Morris.

Extending the Life of Reservoirs • http://dx.doi.org/10.1596/978-1-4648-0838-8

Removal of Sand from Diverted Water

Hydraulic Layout and Short-Circuiting

Sedimentation basins, frequently termed *desanders* or *desilting basins* are used to remove sand from diverted water. These structures are usually constructed as rectangular concrete basins, but underground chambers excavated along the headrace tunnel are also used when space does not allow above-ground construction. These basins are costly to build, especially if underground. They are typically designed to remove sand-size particles exceeding 0.15–0.20 millimeters in diameter; the removal of smaller particles requires larger basins and increases costs.

Efficient sedimentation is achieved when water flows uniformly through a rectangular sedimentation basin along parallel flow paths, and the basin's hydraulic size is normally computed based on this flow condition. In practice, however, the design of sedimentation basins frequently violates one or more of the basic hydraulic principles on which sedimentation theory and computations are based. This results in flow conditions that are far less than ideal, resulting in sand loads and abrasion rates higher than anticipated. These problems are frequently caused by *hydraulic short-circuiting*, meaning that part of the flow and its entrained sediment travels from the basin entrance to the basin exit along a shorter-than-design path, reducing both detention time and sedimentation efficiency, while other parts of the basin are occupied by recirculating flow or dead zones. Problems commonly observed in both old and recently constructed sedimentation basins are summarized below.

- *Flow splitting.* When the diverted water flow is not evenly split among parallel basins, the hydraulic loading rate is increased in one basin and diminished in another. The sedimentation efficiency in the overloaded basin will be diminished and it will discharge oversized and higher-concentration sediment.
- *Entrance jets.* Velocities in the conveyance channel between the intake and the sedimentation basin need to be high enough to prevent sediment deposition, but when this high-velocity flow enters the sedimentation basin it can penetrate as a jet deep into the basin. This jet creates large secondary currents or eddies, including patterns of recirculating flow within the basin and including areas within the basin where water can actually flow back toward the inlet, as illustrated in panel a of figure 8.8. Patterns of surface water recirculation are readily observed by visual inspection, but unseen deeper currents can also be important.
- *Nonparallel flow paths.* Many examples of basin geometry create nonparallel flow paths. The entering flow channel may curve immediately upstream of the basin, thus pushing the flow against one side of the basin as in panel a of figure 8.8. The inlet zone may be nonsymmetrical with respect to the basin as in panel b of figure 8.8. Or instead of having inlet and outlet zones oriented along a straight line, some basins have been designed with exit weirs on the side of the basin instead of the far end. This design attracts the discharge

to one side of the basin while creating a dead zone on the opposite side, as in panel c of figure 8.8.

- *Hydraulic overload.* Because the rate of flow diversion is usually not precisely controlled at the intake, headworks are commonly designed with a weir to discharge excess flow when river levels are high (as shown in figure 8.1). However, in some headworks this overflow weir has been located downstream of the sedimentation basin, passing the excess flow through the basin. This design produces a maximum hydraulic overload in the basin at the worst possible time—during floods when sediment concentrations are highest. To prevent this hydraulic overload, the overflow weir should always be located ahead of the sedimentation basins so that the design flow rate of the basin is never exceeded.

To achieve optimal sedimentation performance in both existing facilities and in new designs it is necessary to minimize the hydraulic problems that produce flow imbalances, hydraulic short-circuiting, and excessive hydraulic loading rates.

Often the least expensive method for enhancing flow patterns is to install a permeable barrier or *flow tranquilizer* in the basin entrance, generating several

Figure 8.8 Undesirable Hydraulic Geometry Observed in Sedimentation Basins

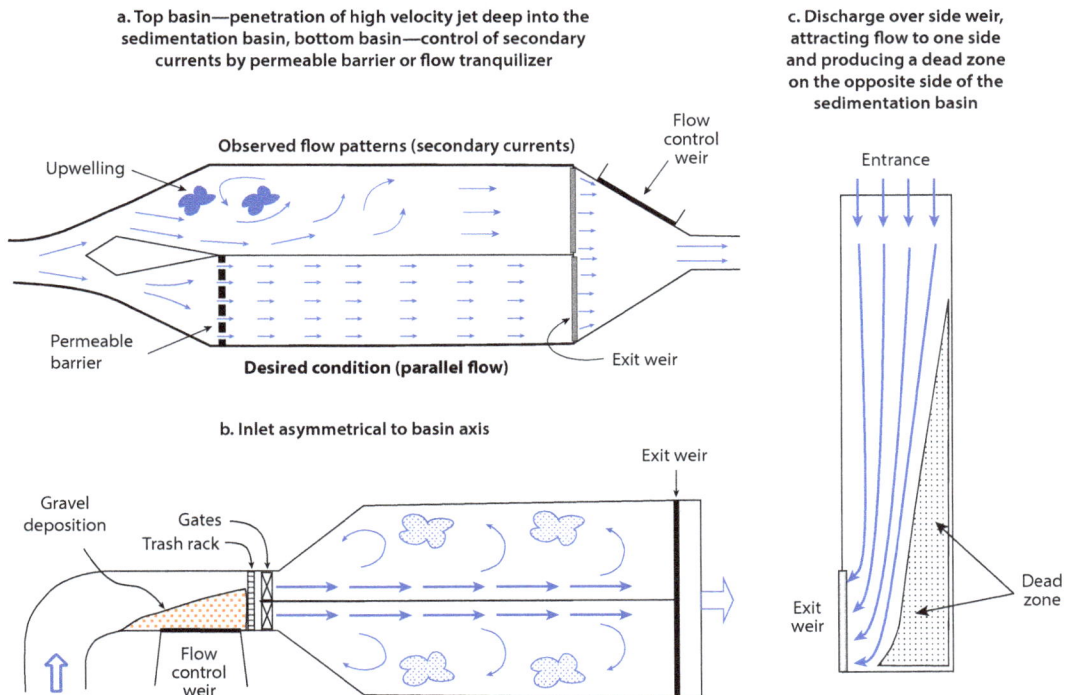

Note: All of the phenomena shown in panels a–c produce hydraulic short-circuiting, which occurs when part of the flow short-circuits to the outlet faster than planned in the sedimentation computations, while other portions of the basin are occupied by recirculating currents or dead zones. This reduces sedimentation efficiency.

centimeters of head loss and breaking the large entrance jet into numerous smaller jets evenly distributed across the cross-section of the sedimentation zone. Design guidelines are given by Bouvard (1992), and a photograph is shown in photo 8.3. In some situations, alternative configurations for outlet weirs may also be installed without excessive expense; for example, by installing a weir across the end of the basin as shown in panel c of figure 8.8 (to replace the side weir) the unfavorable flow path could be straightened and improved.

Proposed basins or proposed modifications to existing basins can be simulated by physical modeling. For existing basins, surface flow patterns can be observed visually, while deeper flow paths can be documented using tracers, current drogues set at different depths, or an acoustic doppler profiler at different cross-sections along the basin length.

Removal of Trapped Sediment

In most sedimentation basins the trapped sediment is removed by periodically emptying and flushing, as shown in photo 8.4. Trapped sediment may also be removed continuously while the basin remains in operation using a series of flushing orifices along the bottom of the basin. Both types of basins normally have steeply sloping (for example, 45 degrees) hopper-type bottoms to cause sediment to slide into the cleanout area. Selection of the cleanout method depends on several factors. Basin emptying and flushing is a simple procedure requiring the minimum level of hydraulic sophistication. However, if the basin rapidly fills with sediment, this will require frequent cleanout and excess sedimentation capacity to compensate for the out-of-service basin. Also, it is not unusual for sediment to be incompletely removed, requiring labor to assist the cleanout. It is not always a completely automatic process.

Photo 8.3 Flow Tranquilizer

Source: © G. Morris. Used with permission. Further permission required for reuse.
Note: Photo illustrates a flow tranquilizer at the inlet to a sedimentation basin showing headloss across this permeable barrier. Typically, two or three rows are used in series.

Photo 8.4 Sedimentation Basin after Emptying for Cleanout, Looking Downstream

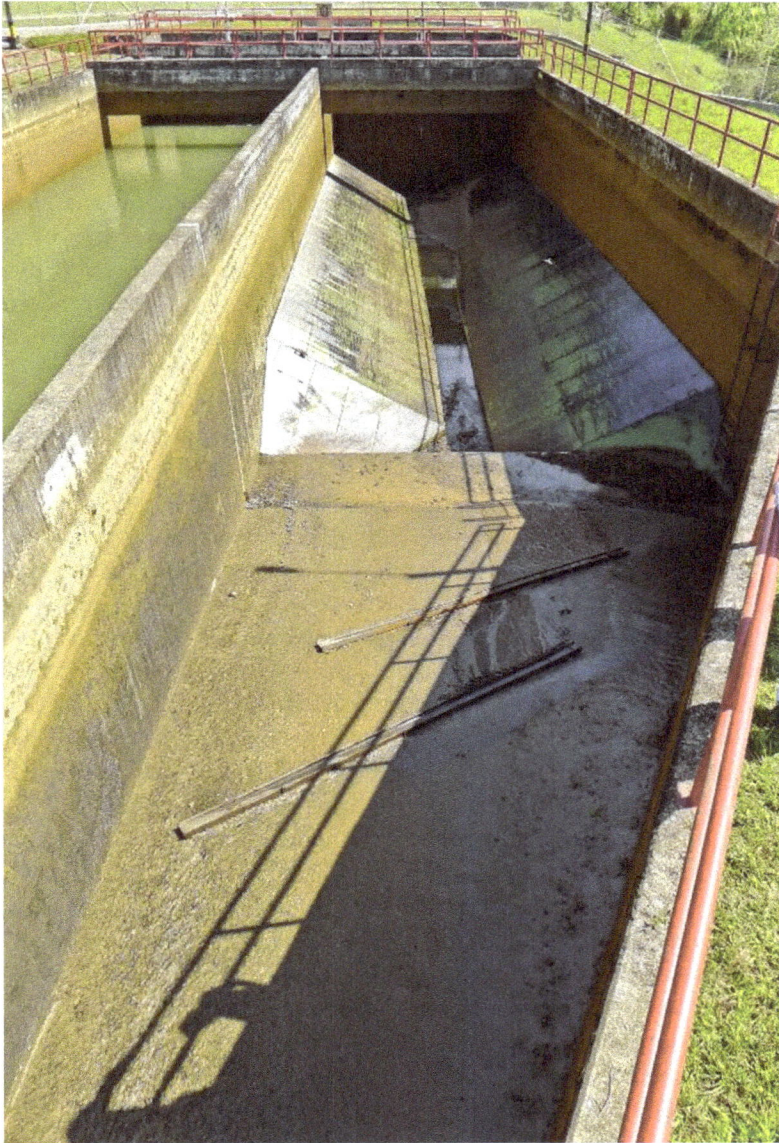

Source: © G. Morris. Used with permission. Further permission required for reuse.
Note: Notice vanes installed on the floor of the expanding basin entrance (foreground) to force cleanout flow across its full width to improve sediment removal. Also notice the hopper bottom in the main sedimentation area.

Continuous removal systems normally use a series of orifices along the bottom of the basin that are opened, one or more at a time, to withdraw sediment from the bottom while the basin remains in operation. Although such systems are more costly and complex to construct, they can facilitate sediment management if the proper conditions are met. Successful operation of this system depends on the sediment slumping and flowing freely into the orifices under submerged conditions, and achieving high flow rates through each of the orifices. These conditions are not always met.

Pure sand will easily slump and flow under submerged conditions. However, if the sediment contains significant fines, the submerged angle of repose may be very steep, preventing the sediment from flowing into the orifice and allowing it to stand up against the sloping side walls, resulting in incomplete removal. Periodically emptying the basin for cleanout then becomes necessary.

In a basin designed for continuous sediment release, an important objective is to have all orifices discharge at the design flow rate. Flow through each orifice or orifice pair may be controlled by an individual valve, but this set-up will increase operational and maintenance complexity, which may not be a wise strategy at many locations. Any valves that do not work, and any orifices that become clogged by debris, will result in incomplete sediment removal. Another alternative is to have multiple orifices discharge into a single undersluice running beneath the bottom of the basin and controlled by a single discharge valve. If this configuration is not properly sized, however, most of the flow will pass through the orifices closest to the undersluice outlet and flow will be greatly reduced through the more distant orifices, resulting in clogging and incomplete cleaning. This problem can be overcome by making undersluice capacity large in relation to the orifices, thereby eliminating orifice backpressure along the length of the undersluice. If the undersluice cross-section is increased, the orifice head loss along the full length of the basin can be equalized, but the velocity may be too low to transport sediment along the upstream end of the undersluice. One strategy is to provide an undersluice that operates under free-flow conditions, with sufficient slope to transport sediment along its full length.

Basins that are continuously cleaned can remain in service during periods of high sediment loading as long as the basin capacity and cleaning cycle are not overloaded by the inflowing sediment load. Basins designed for continuous sediment removal should have a high-capacity flushing outlet to facilitate emptying and hydraulic cleanout when necessary, and easy maintenance access to valves and orifices should also be provided.

Vortex Desander

In rectangular sedimentation basins, the designer seeks to dissipate energy and convert the fast-moving inflow jet into a uniform plug-flow moving uniformly through the basin. In contrast, in a circular vortex desander, the high inlet velocity is used to create a swirling action in which both gravity and centrifugal forces act together to separate sediment from the liquid. Vortex desanders have been used for many decades in grit removal systems, at wastewater treatment plants, for example, but their use in hydropower plants to date has been limited. As an advantage, they may achieve comparable levels of removal with a significantly smaller footprint than conventional rectangular sedimentation basins. Vortex desanders are described by Athar, Kothyari, and Garde (2002), and design guidelines are analyzed by Li (2009). They may be suitable for use in smaller hydropower plants.

Monitoring and Sediment-Guided Operation

Measures of Operational Performance

Headworks should be operated to sustain the reliable diversion of water while minimizing the concentration and grain size of sediment delivered to the turbines. Good or improved performance requires operational records of key performance measures. Such records may include the number of days when water delivery is diminished because of problems at the headworks, the efficiency of sediment removal from the diverted water, inflow and outflow suspended sediment concentration and grain size (outflow concentration can be measured at the power house), water levels in the river to track compliance with the prescribed operating rule, sediment level in the sedimentation basin before each cleanout (for systems without continuous sediment removal), and so on.

For these records to be useful, the information must be displayed in a readily understandable format and not simply buried in a file. Furthermore, the analysis of operational performance data must be regularly reviewed by operational personnel at the headworks as well as their supervisors. If supervisors do not focus on performance records, operators will have scant incentive to collect these data and optimize headworks operation. Operational performance is directly related to the importance that supervisors place on achieving efficient operation, as opposed to simply maintaining the equipment in good condition and similar housekeeping duties.

The efficiency of suspended sediment removal can be assessed by sampling in the river and at the exit of the desanders or the powerhouse. Well-mixed sampling sites are essential for obtaining representative data. River sampling may entail depth-integrated sampling at a river gauge station just above the intake or of the water flowing into the intake.[2] Desander discharge can be sampled at the exit weirs, by depth-integrated sediment sampling in the conveyance channel, or in the draft tubes beneath a Francis turbine or other well-mixed sampling point. The efficiency of sediment removal by desanders may be quantified by comparing inlet and outlet total suspended sand concentration, the concentration above a stated grain size (larger than 0.1 millimeter, for example), or by tracking the d_{84} or d_{90} grain size.[3]

As an example, consider the data plotted in figure 8.9, which computes daily sand removal efficiency of the sedimentation basin by comparing suspended sand concentration in the river at the intake to sand concentration in

Figure 8.9 Decrease in Sediment Removal Efficiency over Time at a Run-of-River Hydropower Plant Correlated to Operator Change in 2010

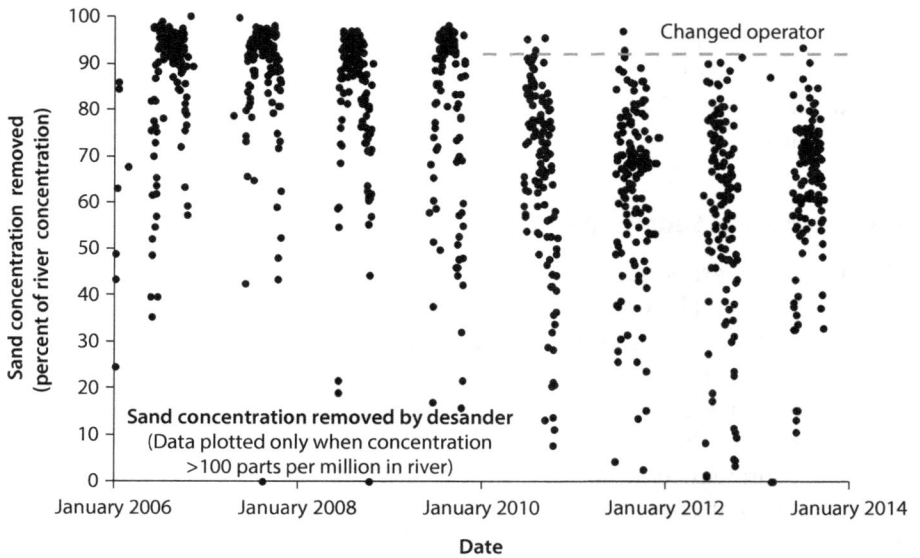

the turbine's draft tube. Notice the significant reduction in sand removal efficiency starting in 2010. This drop was later correlated to a change in operator (less stringent control of water levels in the headpond). No structural modifications were made to the headworks during this period and the operational guidelines were also not modified. These data clearly point to the importance of optimizing operations to achieve the best results. The data also underscore the importance of properly training new operators, tracking performance, and providing operator feedback.

Variation in Turbine Abrasion over Time

Most sediment will be delivered to the turbines during relatively short periods of high streamflows. Even in monsoon climates, with a prolonged period of high summer flow, sediment concentration may be significantly higher at the beginning of the wet season as opposed to the end. Daily suspended sand concentration in the draft tube of a Himalayan ROR plant is plotted as an exceedance graph in figure 8.10, showing that the highest concentration events are very concentrated in time. If the maximum grain size delivered to the turbines also increases with concentration, the abrasion rate will be accelerated because both factors are working simultaneously.

Turbine efficiency data reported at the Jhimruk plant in Nepal show that abrasion reduced turbine efficiency at partial load more than at full load (figure 8.11). This finding implies that the dry season, when energy availability is diminished because ROR plants are running at partial capacity and energy prices are likely to be higher, is also the period when the impact on turbine efficiency will be greatest.

Figure 8.10 Cumulative Daily Sand Load on Turbines at Kali Gandaki Power Plant, Nepal

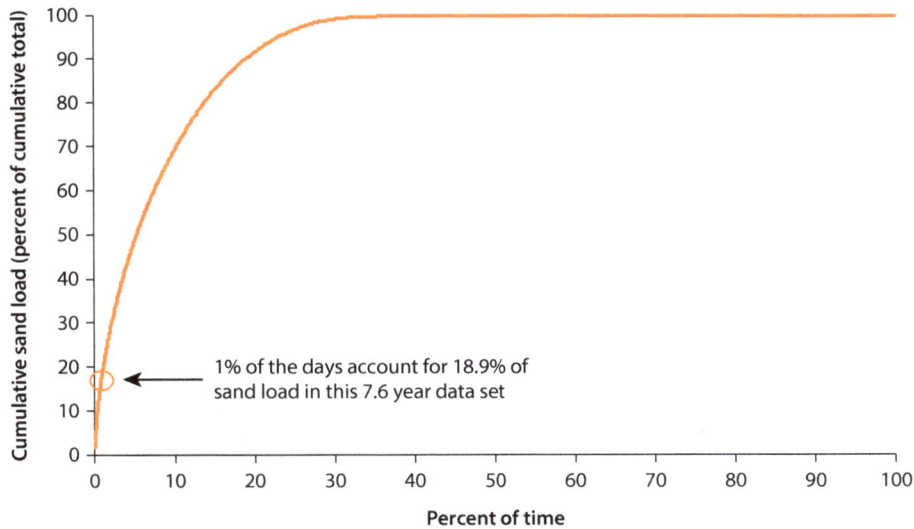

Figure 8.11 Efficiency Measurements at Jhimruk Hydropower Plant, Nepal

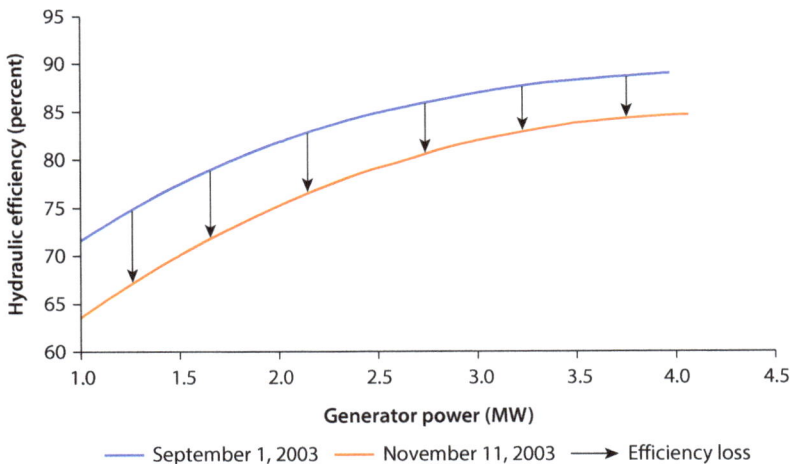

Source: Pradhan 2004.
Note: Figure shows the loss in turbine efficiency due to abrasion. The efficiency declined by 4 percent at the turbine's maximum efficiency point and by 8 percent at 25 percent load. MW = megawatt.

Sediment-Guided Operation

Sediment-guided operation of a hydropower plant refers to the practice of modulating plant operation in accordance with the sediment load with the objective of reducing damage to the equipment. Plant production may be reduced or halted during events that result in high sediment loads. This approach recognizes that there is little advantage in operating a plant during periods when income from power is offset by the cost of equipment damage. This cost is not limited to physical abrasion but also consists of lost power generation due to reduced efficiency caused by the deformation of turbine runner geometry from erosion of the metal. For example, for a turbine operated 180 days per year (50 percent plant factor), each 1 percent loss in efficiency is equivalent to 1.8 days of lost power production each year until the turbine is refurbished.

Sediment-guided operation requires coordination with dispatch to allow power production to be reduced or stopped without adverse consequences to the grid. Sediment-guided operation also requires that sediment concentration at the headworks be monitored in real time, and that plant operation be modified based on these data rather than an established rule based on discharge alone. The highest-concentration events do not necessarily correspond to the highest discharges. Power production can be decreased during periods of high sediment concentration to reduce the hydraulic loading rate on the sedimentation basins, thereby increasing their efficiency, or the plant can be temporarily shut down to avoid damage.

Real-Time Data Collection

Real-time monitoring of suspended sediment concentration and particle size distribution can be performed either manually or automatically. A rapid manual method for measuring high sediment concentration in water samples would be to use a specific gravity bottle (pycnometer). The size distribution of sands can be rapidly measured manually using a visual accumulation tube, filtering flow through a screen to obtain a sufficient volume of sand to run the test. Automated curves of particle size distribution up to 0.5 millimeter, and the calculated concentration, can be continuously measured in real time by laser diffraction. These instruments are now available in field-deployable models suitable for use in hydropower plants (for example, the LISST instruments from Sequoia Scientific).

Particular care should be given to the selection of sampling location and techniques to ensure the sample is representative of the entire flow. Surface grab samples from a conveyance channel will underestimate sand concentration unless the location has sufficient turbulence to be completely mixed. If sampling is occurring in the headrace channel following the sedimentation basins, a depth-integrated sampler should be used for sample collection.

A relationship between sediment and plant characteristics and abrasion rates was developed by Nozaki (1990) based on average conditions. However, the collection of data sets from instruments now being placed into the field can be used to observe the variation in abrasion parameters over relatively short time steps. From such data it may be possible to develop a better empirical understanding of the relationship between time-variant sediment characteristics and abrasion rate, and to help develop sediment-guided operating rules to improve plant efficiency and reduce operational expense.

Notes

1. To better comprehend the nature of these flows, the reader is encouraged to perform an Internet search for "debris flow boulders" and view several videos.

2. Sampling performed at the intake entrance will not necessarily be representative of the river as a whole, but it will be useful for computing desanding efficiency.

3. The d_{84} and d_{90} size refer to the grain size that 84 percent or 90 percent of the sampled particles are smaller than, as shown on a particle size distribution curve.

References

ASCE (American Society of Civil Engineers). 1995. *Guidelines for Design of Intakes for Hydroelectric Plants.* New York: American Society of Civil Engineers.

Athar, M., U. C. Kothyari, and R. J. Garde. 2002. "Sediment Removal Efficiency of Vortex Chamber Type Sediment Extractor." *Journal of Hydraulic Engineering* 128 (12): 1051–59.

Bouvard, Maurice. 1992. *Mobile Barrages and Intakes on Sediment Transporting Rivers.* Rotterdam: A.A. Balkema.

Li, Yunjie. 2009. "Development of Design Basis for Hydrodynamic Vortex Separators." Rutgers, The State University of New Jersey, New Brunswick, NJ.

Nozaki, Tsuguo. 1990. "Estimation of Repair Cycle of Turbine Due to Abrasion Caused by Suspended Sand and Determination of Desilting Basin Capacity." Japan International Cooperation Agency, Lima, Peru.

Pradhan, Pratik Man Singh. 2004. *Improving Sediment Handling in the Himalayas.* Nepal: OSH Research.

Rosgen, D. L. 1994. "A Classification of Natural Rivers." *CATENA* 22: 169–99.

———. 1996. *Applied River Morphology.* Pagosa Springs, CO: Wildland Hydrology Books.

CHAPTER 9

Reservoir Sustainability Best Practices Guidance

Gregory L. Morris

Introduction

The objective of sustainable sediment management in reservoirs may be expressed as follows: *Sustainable sediment management seeks to maintain long-term reservoir capacity, retarding the rate of storage loss and eventually bringing sediment inflow and discharge into balance while maximizing usable storage capacity, hydropower production, or other benefits.*

Sustainability is an overarching theme of the World Bank Group, and this sustainable development definition is consistent with the World Bank's stated objectives:

> Sustainable development recognizes that growth must be both inclusive and environmentally sound to reduce poverty and build shared prosperity for today's population and to continue to meet the needs of future generations. It must be efficient with resources and carefully planned to deliver immediate and long-term benefits for people, planet, and prosperity.[1]

This chapter summarizes sediment management strategies that are central to the sustainable design and management of dams and reservoirs.

Creating and maintaining reservoir storage volume is required to convert irregular stream flows into reliable supplies of fresh water for agricultural, domestic, and industrial use; for flood management; and to maximize hydropower production. Groundwater is globally overexploited and in most areas cannot compensate for declining water supplies caused by reservoir sedimentation. Dams have been constructed for a wide variety of purposes (figure 9.1), and all but a limited number of run-of-river hydropower dams have in common the need to sustain storage to deliver benefits. In all cases, including run-of-river hydropower, project function cannot be sustained indefinitely without sediment management.

Of particular concern are the 64 percent of dams constructed for irrigation, water supply, and flood control that directly depend on reservoir storage for

Figure 9.1 Designated Beneficial Uses of Reservoirs Worldwide
(percentage of total number of reservoirs reporting)

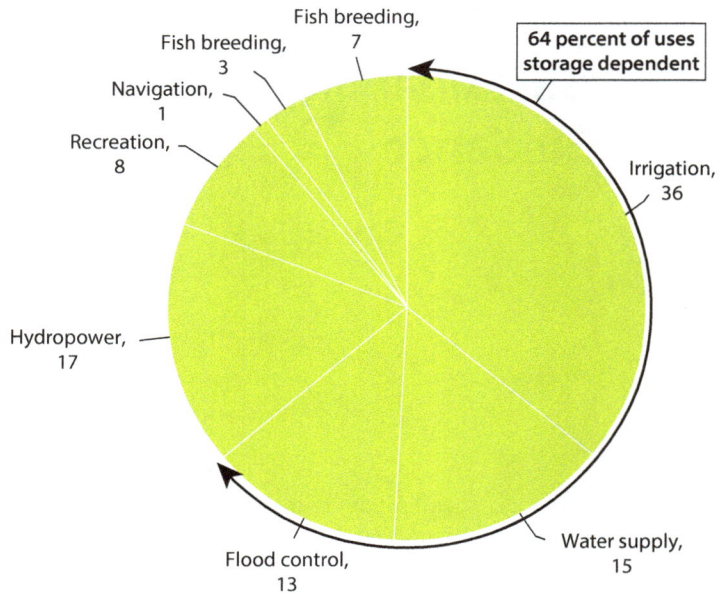

Fish breeding, 7
Fish breeding, 3
Navigation, 1
Recreation, 8
64 percent of uses storage dependent
Irrigation, 36
Hydropower, 17
Water supply, 15
Flood control, 13

Source: Data from ICOLD 2015.

streamflow regulation to produce their benefits; as storage is lost these benefits also disappear. A significant fraction of hydropower reservoirs also provide streamflow regulation, and conversion from storage to run-of-river operation because of sedimentation may reduce both energy production and power peaking potential. The reservoir volume required for streamflow regulation is a limited resource that will be irrevocably lost in the absence of sediment management. The consequences of declining storage volumes in storage-dependent projects will become more acute to the extent that hydrologic variability increases (more severe floods and droughts), an expected consequence of climate change.

All hydropower projects, including today's large storage projects, will eventually require sediment management to sustain operation. Hydropower is a uniquely long-lived renewable resource, and a number of hydropower plants have now been in operation for more than 100 years. This long-term availability of renewable energy is an important benefit of hydropower development, and proper sediment management is key to achieving these results.

Sustainable Reservoirs and Hydropower

Sustainable Approach to Sediment Management

Dam engineering has traditionally been based on the "design life" or "life of reservoir" paradigm, which typically provides sufficient inactive pool volume to store 50 or 100 years of sediment but which does not consider consequences beyond this design life. Sustainable sediment management is a new and contrasting paradigm that focuses on managing the reservoir and watershed system to bring sediment inflow and outflow into balance insofar as is practical, thereby giving the reservoir a greatly extended or even indefinite lifetime. The sustainable management paradigm may be applied to new projects as well as to existing projects, as conceptually illustrated in figure 9.2. It is a linear concept that extends reservoir life far into the future by applying specific sustainability interventions. These interventions will typically be applied as a sequence of actions undertaken over time as sedimentation progresses.

Sustainable use does not mean that projects will continue to deliver the same benefits they did when they were new; the permanent loss of significant storage capacity will often be inevitable and storage recovery too costly to undertake on a significant scale. This is particularly true of the largest storage reservoirs, some of which are of extremely high national importance such as the Arab Republic of Egypt's Aswan Dam on the Nile. Rather, sustainable use implies that existing reservoirs can, through modification, successfully move from their initial operating configurations, which continuously trap sediments, into sustainable operation that more closely balances sediment inflow and discharge, while continuing to generate significant benefits. New dams and reservoirs may be designed for long-term sustainable use from the beginning. Sustainability concepts as applied to reservoirs are discussed by Morris and Fan (1998); Palmieri et al. (2003); Annandale (2013); and others.

Numerous factors influence the viability of implementing a sustainable use strategy and the selection of specific techniques. Major factors requiring consideration are shown in figure 9.3. Environmental, economic, and social factors are just as important as engineering factors in defining viable sediment management alternatives.

Figure 9.2 Contrasting Design Life and Sustainable Use Paradigms

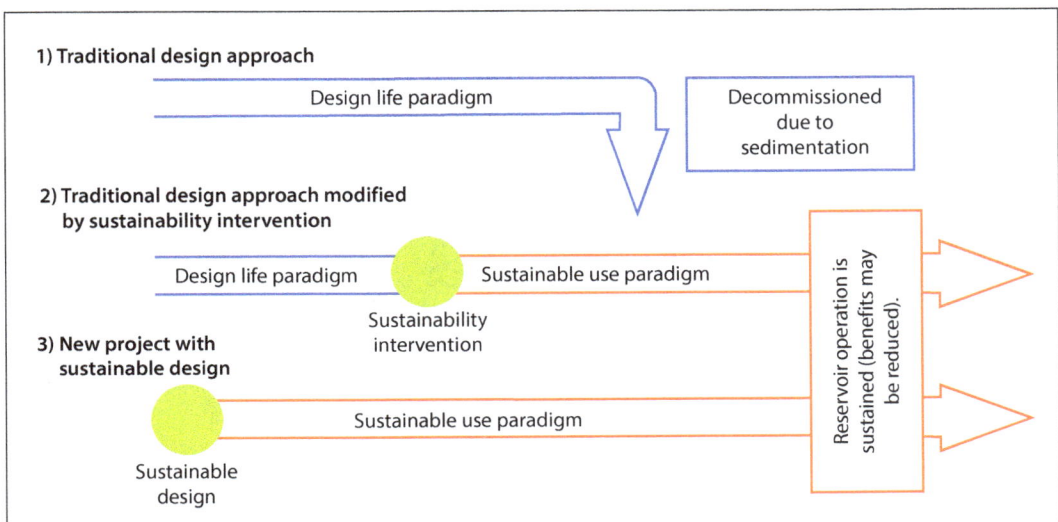

Figure 9.3 Major Factors Influencing Sustainable Use Strategies

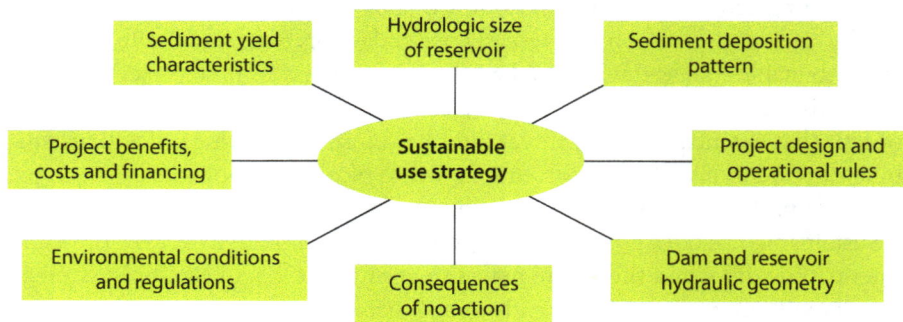

The hydrologic size of the reservoir, defined as the ratio of reservoir volume to mean annual inflow (the capacity-inflow ratio), is a particularly important factor in determining the type of sediment management strategies that can be applied (recall figure 7.14). Hydrologically small reservoirs that frequently discharge flood flows have excess water available for the routing of sediment-laden floods or sediment flushing. However, in hydrologically large reservoirs, especially those having more than a year of storage capacity to capture and regulate all inflow, the only viable strategy for the long-term maintenance of storage is to reduce sediment inflow and implement density current venting where feasible. While this strategy will reduce the rate of storage loss, there is currently no economically viable method to balance sediment inflow and discharge to sustain such large storage volumes indefinitely. As storage volume is lost and the large reservoir becomes a smaller reservoir because of sedimentation, new management options for releasing sediment to retard the rate of storage loss may then become viable.

Sustainability Considerations for BOT Projects

Build-operate-transfer (BOT) projects are becoming increasingly common in the hydropower sector. In BOT projects, the project developer builds and then operates the project for a period of decades, achieving a return on the developer's investment, and then transfers the operating project to the contracting entity, such as a national electric company. The contracting entity wants to take control of the project in good working order with the expectation of many additional decades of trouble-free operation. However, BOT operators do not have an intrinsic interest in sediment management if it entails near-term costs that produce benefits beyond the term of the BOT contract. By incorporating sediment management requirements into the BOT contract the owner can receive a more sustainable project at the end of the contract term.

The sustainable operation proposal advanced by the proponent of a BOT project should be subject to the same level of scrutiny that it would were the contracting entity to construct the project with its own funds. To achieve a sustainable project, the contracting entity must require sediment management measures of the BOT proponent, and the contracting entity's project review team must have a high level of competence in this field. The project sustainability review should focus, in particular, on sediment management measures and any required modification to the operational rule at least 100 years into the future. In the event that the project operation cannot be sustained indefinitely, an end-of-life scenario that is acceptable from the standpoint of dam safety, environmental impact, and implementation cost needs to be clearly defined. It should specifically address the question of what will happen when coarse bed load sediment reaches the dam and must be passed downstream.

Limitations of Sediment Management

Volume Loss in Storage Reservoirs

As pointed out in chapter 2, the need to sustain storage volume is a critical issue globally. Irrigation is the largest consumptive use of water worldwide, and more reservoirs are used for irrigation supply than any other use (see figure 9.1). Maintaining water supplies for irrigation is a critical need that is threatened by sedimentation of reservoir storage. The challenge facing irrigation reservoirs is particularly acute given their limited potential to generate the income needed to cover sediment management costs.

Achieving a sediment balance, or even discharging significant amounts of sediment, will not be feasible at many storage reservoirs. Hydrologically large reservoirs having a capacity exceeding 0.5 times mean annual inflow face the greatest problem. This size is generally the absolute upper limit for implementation of hydraulic management procedures such as sediment flushing, sluicing, and bypassing (see figure 7.14). However, these techniques are normally used, and are more feasible, in much smaller reservoirs that typically have capacities of not more than about 0.1 mean annual flow. The most feasible approach to sediment release for larger reservoirs is to discharge turbid density currents, but this strategy will often be ineffective because turbidity currents do not transport a large amount of sediment to the area of the dam in many reservoirs.

Once sedimented, the potential to recover reservoir capacity is very limited. Flushing has been practiced to recover capacity at some hydrologically small reservoirs, but this strategy has significant technical limitations. As pointed out in figure 7.13, reservoir emptying will only scour and recover capacity along a narrow flushing channel. Also, unacceptably large volumes of water may be required to remove sediment, especially in the case of cohesive deposits. To this is added the problem posed by discharging large volumes of sediment downstream in a manner acceptable to both downstream users (including downstream dams) and the aquatic environment.

Dredging is technically feasible at virtually all reservoirs, but economically feasible at relatively few sites because of high cost together with limited long-term disposal options. While the number of reservoir dredging projects is certain to increase in the future, to date it has only infrequently been an economically attractive option.

Extending the Life of Reservoirs • http://dx.doi.org/10.1596/978-1-4648-0838-8

Because of these limitations, it should not be assumed that sediment management can be implemented in the future to stop sedimentation, or that recovering lost capacity will be feasible. While sediment management can be very successful under the appropriate conditions, and when undertaken in a timely manner, it is not a miracle cure for all the ills of sedimentation; therefore,

- Sediment monitoring and management for sustainable use should start as soon as possible and be incorporated into the original designs for new projects.
- At hydrologically large reservoirs particular attention must be given to reducing the rate of sediment inflow, since other options to preserve large reservoir capacity—except possibly turbidity current venting—will probably not be feasible until most of the storage volume has been lost.

The longer that sediment management is delayed, the worse will be the severity of the problem, and the solutions will become more costly. Sustainable sediment management strategies should be developed for all reservoirs and be accompanied by appropriate monitoring. Future structural modifications and management activities should be aligned with the long-term sustainability strategy. A screening analysis can be undertaken to identify those sites having the highest priority for in-depth analysis and remedial action. Management strategies should be periodically revisited and revised in light of additional monitoring data plus changes in technology, costs, and benefits.

Finally, addressing the problem of sedimentation within the larger water management context is essential. Sediment management techniques such as dredging may be much costlier than options such as the development of an improved real-time operational rule to better use reduced storage volume, the conjunctive use of multiple resources including groundwater, or even the construction of a replacement reservoir. For example, because the water storage function of Welbedacht Dam (figure 5.9) was being lost to sedimentation, the nearby Knellpoort Dam was placed into operation in 1989. To control sedimentation, the new Knellpoort Dam was located offstream (panel b of figure 7.4), and Welbedacht Dam is operated to maintain only a pool in front of the water supply intake (de Villiers and Basson 2007; Basson 2015 [personal communication]).

Volume Loss in Hydropower Reservoirs

Irreversible storage loss in hydropower reservoirs is a less critical problem from the standpoint of long-term sustainability of our society than is storage loss in water supply reservoirs. While there is no substitute for water, there are substitute sources of renewable energy, particularly given the steadily declining cost of photovoltaics and energy storage systems (in addition to pumped storage hydropower, which currently provides more than 95 percent of utility-scale power storage). From the standpoint of the hydropower operator, even if storage volume is lost the hydropower plant can continue to produce a large amount of energy operating in run-of-river mode. Furthermore, if even a small percentage of the original volume can be preserved, not only can it be used for trapping sediment to protect turbines from abrasion, but it can also sustain the volume required for daily or multiday power peaking storage, to be operated in conjunction with intermittent renewable energy sources from solar and wind. The growth of renewable energy has increased the importance of hydropower for power peaking. The transition of a hydropower storage project to a sustainable run-of-river or power peaking operation will typically require both operational and structural modifications. Experience to date indicates that many storage hydropower projects can make this transition at a reasonable cost and continue to operate economically, especially given the potentially high value of storage for peaking power (recall table 8.2).

Planning and Design Considerations

To design a new reservoir for sustainable use requires that the designer consider conditions beyond the financing period, the BOT concession, or the nominal design life. Otherwise, the project runs the risk of working satisfactorily for only a few decades before encountering sediment problems that severely restrict its benefits, that require large expenditures to remedy, or that result in project abandonment and potentially large decommissioning costs. To achieve sustainable operation the following design requirements should be considered for new projects:

Hydrologic Data Collection

The value of accurate long-term discharge data sets collected by governmental institutions and made publicly available cannot be overestimated. Ideally the discharge data set for the project river will be complemented by long-term data sets collected from other rivers in the region. Regional data should be used if needed to extend the available data set, creating a discharge data set that is as long as possible. This discharge time series may be combined with a sediment rating relationship to estimate long-term sediment yield and its variability.

Suspended Sediment Field Data Collection

Field data on suspended sediment concentration are essential for construction of sediment rating relationships. Suspended sediment data collected over several years may be combined with much longer discharge data sets to estimate sediment yield. For sites with predictable seasonal flows (e.g. monsoons) and very little dry-season sediment transport, data collection programs should particularly focus on the wet season.

If suspended sediment data are not available, a sampling program should be implemented to obtain data for several years to construct a sediment rating relationship, compute long-term sediment yields, and analyze both seasonal and annual variability in sediment load. These parameters can be important in identifying appropriate sustainability measures. Field data should include information on the grain size of both the suspended sediment and the bed material. For short data sets, field data collection should continue throughout the duration of the planning and design phase to further verify the design data. Approximate methods, including transposition of data from other watersheds and other empirical techniques presented in chapter 4, may be used for a preliminary analysis and to validate the results of more detailed studies, but should not be used as the basis for project design. Projects should not be undertaken without field data to verify the accuracy of any sediment yield assumptions incorporated into the design.

Sediment Yield

Data for suspended sediment concentration and river discharge are used to construct a sediment rating curve. Applying this curve to a longer hydrologic data set of daily flows provides the best means of estimating long-term sediment yield and variability. Errors frequently occur in developing the sediment rating relationship. Particular attention should be paid to the form and accuracy of rating curves (see the "Sediment Rating Curves" section in chapter 6), and to the accuracy and adequacy of the data on which they are based.

Because of the high variability in year-to-year discharge in some rivers, the potential for considerable error is present when using short data sets that do not include high discharge events. Although sediment rating relationships will change from year to year, they tend to be much more stable than the year-to-year variability in yield resulting from the variability in discharge. At least four years of suspended sediment for the rating curve is recommended, and more data can be very important when they capture high discharge events. Thirty or more years of daily discharge data is also recommended. Use of 10-day, monthly, or similar averaged values should never be substituted for daily values when computing sediment yield from a rating relationship.

The sediment yield estimate should be compared with specific sediment yield data (tons per square kilometer per year) derived from reservoir surveys, other gauge data in the same physiographic region, and results from empirical methods (see the "Sediment Yield Estimation" section in chapter 4). Data points from different sources should all be plotted on a log-log graph of sediment yield vs. drainage area to better visualize the range of sediment yield in the region and the reasonableness of the selected design value. Long-term regional data sets should also be examined to better understand the timewise variability in sediment yield. Sediment yield variability over time may be particularly high in mountainous and semi-arid regions.

Extreme Sediment Events

Many areas experience infrequent but extreme sediment discharge events, including debris flows, glacial lake outburst flood events, and extreme floods. These events may have a recurrence interval exceeding 100 years and are rarely captured in short-term suspended sediment data sets. However, they can have catastrophic consequences for a project (recall photo 8.1). Particular attention should be given to determining the hazard posed by such events through review of historical records in the region and geomorphic analysis of the river and its watersheds. Project designs should focus on minimizing damage to facilities and enabling rapid recovery from such events.

Data Sets

All quality-controlled (corrected) original data sets used in the analysis should be included with project documentation in electronic format for project review and for use by subsequent investigators (for example, spreadsheet data files can be embedded in the project report's PDF file). Daily concentration-discharge data points should be included as a printed appendix in the sedimentation report in the event that electronic files are corrupted, unless they are available through an established national data repository.

Bed Load

Measuring bed load transport is not feasible on most rivers. Bed load transport is usually estimated from equations (for example, Parker 1990; Wilcock and Crowe 2003), by applying data from other reservoirs, or as rule-of-thumb estimates (see the "Bed Load Estimation" section in chapter 4). The bed material grain size should be documented by sampling at several locations along the river at low flow, including documentary photographs and global positioning system coordinates. The method for estimating bed load should be clearly stated and a sample calculation given.

Sedimentation Modeling

Long-term reservoir sedimentation should be simulated by modeling, extending the modeling period until a stable longitudinal profile and sediment balance across the dam have been achieved. One-dimensional models are currently considered appropriate for long-term simulations, but two-dimensional and physical modeling may also be appropriate in some cases. Use of empirical techniques, such as the area-reduction method or van Rijn's (2013) method may be useful at the prefeasibility stage, but should not be used for design purposes because the techniques do not adequately incorporate site-specific characteristics such as sediment grain size, hydrology, reservoir geometry, and operating rule. Results from modeling provide multiple benefits.

- The modeling exercise helps develop and formalize both the qualitative and quantitative aspects of sediment management.
- Modeling provides a better idea of the long-term future conditions anticipated in the reservoir, including conditions at the time of a BOT project transfer.
- Modeling assists in the evaluation and quantification of the effectiveness of operational rules for sediment management. It will provide an idea of the extent and geometry of sediment deposits, which will help designers select the location and configuration of outlet works to best handle anticipated future sedimentation conditions. (This phase of the analysis will typically also require physical modeling.)

Although the precise timing and pattern of sedimentation will necessarily be uncertain, modeling can nevertheless help identify the preferred types of both short- and long-term sediment management strategies, their implementation sequence, and the structural measures required to support sustainable operation.

Modeling only provides an approximation of sediment transport conditions, and model results should never be blindly accepted. All modeling results should be carefully evaluated with respect to their reasonableness and interpreted appropriately.

For BOT projects in particular, it is important that the software model used for analysis be publicly available (as opposed to "proprietary" software). The contracting entity should be provided an electronic copy of the fully functional model for the purposes of reviewing results and analyzing additional scenarios.

Downstream Impacts

Analyzing how dam construction will modify the sediment transport regime below the dam is important because dams can have major impacts on the downstream river, including bed incision and accelerated bank erosion. As a mitigating measure, flood hydrographs below most storage dams are significantly reduced, even if the dam is not designated as a flood control structure. Thus, at the same time that the dam is trapping the coarse bed material, the postdam hydrology may reduce below-dam transport capacity, thereby mitigating downstream scour to at least a limited degree. The following types of activities may be undertaken to predict and monitor downstream impacts:

- *Measure cross-sections at selected locations and characterize the bed material extending many kilometers downstream of the dam.* The distance will vary depending on local conditions, but investigations extending more than 100 kilometers downstream may be appropriate along alluvial rivers with limited lateral sediment inflows. These data are required to determine downstream susceptibility to scour and bed degradation, and to provide baseline data for monitoring during the operational phase. Because changes will naturally occur along alluvial rivers even absent upstream disturbance, documentation of natural (preproject) patterns of river behavior from historical data, aerial photography, and prior studies can be important.
- *Predict the postdam discharge time series below the dam to provide input data for the sediment transport model of the downstream reach.* Simulating the routing of floods through the reservoir will normally be necessary to produce the below-dam discharge time series and flood hydrographs. Evaluate the potential for long-term channel incision and change in particle size distribution below the dam.
- *Inventory downstream users and environmental resources* that may be affected by a cutoff in the bed material supply resulting from dam construction, or that may be sensitive to planned management operations such as daily flow changes from hydropower peaking (which can accelerate bank erosion). Structures particularly sensitive to the postdam sediment regime, such as bridges, may require special attention. Incision of the downstream streambed can lead to bridge failure by scour and increased rates of bank erosion.

Extending the Life of Reservoirs • http://dx.doi.org/10.1596/978-1-4648-0838-8

High suspended sediment concentrations associated with reservoir flushing may affect downstream uses including municipal water supplies, irrigation diversions, cooling water intakes, fisheries, and recreation in addition to ecosystem impacts.

Downstream impacts will vary with site conditions, but are typically larger for project sites located further downstream in the watershed; mountain streams tend to have lower ecological diversity than larger downstream rivers, and mountain streams will frequently have bedrock controls that limit incision. However, mountain rivers are often the principal source of both suspended and bed material supplying downstream alluvial reaches. Thus, a mountain reservoir that cuts off the downstream supply of sediment, and particularly the supply of bed material, can have long-term consequences on the downstream river channel and ecosystems, even though the project itself may be constructed in an area of reduced environmental sensitivity.

Major sediment-related downstream environmental impacts due to the trapping of coarse sediment include modification of downstream channel morphology and degradation of spawning habitat. The trapping of fine sediment and particulate organic matter will interrupt food and nutrient flows along the river. Water storage will affect the natural hydroperiod and may reduce or eliminate flood peaks responsible for mobilizing and cleansing accumulating fine sediment from river gravels used for spawning or as habitat for aquatic invertebrates. Monitoring strategies for downstream aquatic ecosystems and associated riparian habitats must be designed on a case-by-case basis.

Upstream Impacts

Long-term sedimentation impacts above the normal pool elevation should be considered, since delta deposition can cause both riverbed and water levels to rise upstream of the reservoir. Impacts may include impediments to navigation as the stream changes from a single deep channel to a shallow or braided channel flowing across the delta, increased flood levels, waterlogging of soils, reduced bridge clearance due to sedimentation, and sedimentation of intakes or irrigation diversions.

Monitoring Sediment Management Performance

All projects should be monitored for sedimentation during the operational phase, with the type and frequency of monitoring dependent on project characteristics. For storage reservoirs, including run-of-river projects with pondage, monitoring should generally include the following types of data:

- *Bathymetric surveys* should be performed soon after initial filling; after 5, 10, and 15 years of operation; and thereafter as appropriate to the site (for example, document every 5 percent decline in pool volume). At projects with high rates of sedimentation, a survey interval of one or two years may be appropriate. Surveys must be performed using a consistent methodology as previously discussed in the "Bathymetric Mapping of Sedimentation" section in chapter 6. Results should be presented as storage-elevation and storage-area curves, a longitudinal profile, and selected cross-sections, with each additional set of survey data superimposed on the previous data (for example, see figures 6.5, 6.6, and 7.9).

- *Sediment cores* should be taken or delta sediments should be sampled during drawdown to determine particle size distribution and bulk density at selected locations. For BOT projects, sampling should also be performed during the last year of the concession. In some reservoirs the delta sediment will be too weak to support field crews (mud, quicksand) and there may be ponding water. In these cases, sampling along the more accessible banks of the river channel crossing the delta will represent the coarser fraction of the deposited delta sediment. To obtain more representative sediment samples on deltas with large areas inaccessible by foot, performing sampling of submerged sediment once the delta is submerged may be a better approach. Multiple samples across the width of the delta, at several cross-sections, can better represent the overall grain size distribution of the delta deposit as compared to channel-bank samples only.

- The *suspended sediment concentration and particle size distribution* of water delivered to hydropower turbines or other outlet works should be characterized. At storage projects this may include sediment delivered by turbid density currents and sediment scoured from the delta by flood flows during drawdown. At run-of-river projects, daily data (including near-continuous data during high flow periods) may be needed to better characterize the changing sediment load on the equipment and to monitor the performance of desanding operations during seasonally high flows.

River cross-sections and documentation of bed material grain size at fixed locations below the dam may also be required to monitor geomorphic changes due to dam construction and operation. Stream gauging may also be needed to monitor operational impacts and the effectiveness of sediment management activities such as sluicing or flushing. Additional sediment monitoring concepts are presented by Morris (2015).

End-of-Life Scenarios

Addressing end-of-life scenarios for dams and reservoirs is important. At storage hydropower projects a variety of options are available to achieve the economical transition to run-of-river operation after the storage volume has become seriously depleted, thereby avoiding an end-of-life scenario. However, options for water supply storage reservoirs are much more limited, particularly in the irrigation sector, which has less capacity to pay for sediment management than do the municipal and industrial sectors.

The end-of-life scenario may require project decommissioning with its attendant costs and impacts. As an alternative, before the storage capacity has been seriously depleted the project's structures and operating rule may be modified to achieve a sediment balance and sustainable operation. The key to sustaining operation is to identify the desired outcome and management strategy as early as possible, and to implement the required modifications while there is still time to save a significant portion of the storage volume.

Note

1. "Overview" (http://www.worldbank.org/en/topic/sustainabledevelopment/overview).

References

Annandale, G. 2013. *Quenching the Thirst: Sustainable Water Supply and Climate Change.* Charleston, SC: CreateSpace.

de Villiers, J. W. L., and G. R. Basson. 2007. "Modeling of Long-Term Sedimentation at Welbedacht Reservoir, South Africa." *Journal of the South African Institution of Civil Engineering* 49 (4): 10–18.

ICOLD (International Commission on Large Dams). 2015. "Register of Dams." http://www.icold-cigb.org/GB/World_register/general_synthesis.asp.

Morris, G. L. 2015. "Collection and Interpretation of Reservoir Data to Support Sustainable Use." SEDHYD 2015, 10th Federal Interagency Sedimentation Conference, Reno, NV, April 19–23.

Morris, G. L., and J. Fan. 1998. *Reservoir Sedimentation Handbook.* New York: McGraw-Hill.

Palmieri, A., F. Shah, G. W. Annandale, and A. Dinar. 2003. *Reservoir Conservation: The RESCON Approach.* Washington, DC: World Bank.

Parker, G. 1990. "Surface-Based Bedload Transport Relation for Gravel Rivers." *Journal of Hydraulic Research* 28 (4): 417–36.

van Rijn, L. C. 2013. "Sedimentation of Sand and Mud in Reservoirs in Rivers" (accessed June 26, 2015). http://www.leovanrijn-sediment.com/.

Wilcock, P. R., and J. C. Crowe. 2003. "Surface-Based Transport Model for Mixed-Size Sediment." *Journal of Hydraulic Engineering* 129 (2): 120–28.

Checklist for Sediment Management

This appendix presents a checklist based on sediment problems that have been important in different projects. Each checklist question should be assessed to determine whether it is applicable to the project under consideration and whether it should be evaluated in greater detail. This list is not all inclusive; other problems may be important depending on site conditions.

Sediment Yield

If fluvial sediment data are not available, how has sediment yield been estimated?
Sediment data may not yet be available during the initial project planning phase, and in such cases it may be necessary to make a preliminary estimate of sediment yield using empirical techniques (see "Sediment Yield Estimation" section of chapter 4).

☐ What methods were used to develop empirical estimates? Refer to the "Empirical Methods" section of chapter 4.

☐ Have different methods been used, and by how much do their results differ? Averaging such values may be common practice, but it is not necessarily desirable and does not represent a "conservative" approach at the project planning stage.

☐ Can sediment yield be expected to change significantly in the future as the result of anthropogenic impacts, glacial retreat, and other factors? Do the empirical methods account for future changes? Refer to the "Empirical Techniques" section in chapter 4.

☐ Has the empirical model been verified against depth-integrated sediment data or reservoir survey data at other similar locations?

If fluvial sediment data are available, how accurate and representative are they?
Ask the following questions to reduce some of the common sources of error that frequently affect sampling technique, laboratory technique, and mathematical curve-fitting of sediment rating equations.

☐ Have suspended sediment sampling procedures been reviewed to assess the reliability of the data? Were the samples collected using depth-integrated sampling by institutions with recognized quality standards? If not, then additional questions must be asked to determine the accuracy and consistency of sampling results, and a site visit to observe and check sampling and laboratory technique is recommended. Water quality data, usually collected as surface grab samples, should not be used to estimate suspended sediment load.

☐ Has the rating curve been validated by back-testing against the data from which it was developed? Refer to the "Sediment Rating Curves" section in chapter 6 and to figure 6.4.

☐ Does the sediment sampling period include accurate sampling of high-discharge events responsible for large amounts of sediment delivery? If not, what assumption has been made to extend the rating curve beyond the available data? To what degree does this assumption influence the computed sediment yield?

☐ Has the specific sediment yield estimate (in tons per square kilometer per year) been verified by comparison with gauge data from other streams in the physiographic region, with measured rates of sedimentation in existing reservoirs, or with empirical models?

☐ Has the particle size distribution of the suspended load been measured on multiple occasions covering a wide range of discharges using a depth-integrated sampler?

How has bed material load been estimated?
In addressing the following questions, bed material load is defined as that portion of the total load that is not measured by suspended sediment sampling.

☐ Has the particle size distribution of the bed material been quantified by field measurement at multiple locations in the reach above the proposed reservoir?

☐ Is the coarse bed material load considered to be a significant contributor to overall sediment load? How has the bed material load been determined? Was more than one method used to estimate bed load and if so are the methods in agreement?

☐ As a point of comparison, has the accumulation of coarse bed material in existing reservoirs in the region been evaluated, and are observations at these reservoirs consistent with the predicted transport rate?

Are there major uncertainties that may influence future sediment yield?

Future conditions may differ from past conditions, and infrequent but large hydrologic events may not be represented in the available data set. Land use change is the largest factor affecting changes in sediment yield over time. Climate change can also contribute to changes in sediment yield, such as in high mountain areas where glacial retreat uncovers sediment beds, where snowfall may have changed to rainfall, or where storm intensity increases, all of which can increase sediment yield.

☐ Has the watershed been analyzed for major episodic sediment transport hazards, which may include debris flows, landslides, and glacial lake outburst flood events?

☐ Is the watershed susceptible to processes that may cause sediment yield to change over time, such as changes in land use or climate change?

☐ Is it possible to estimate the direction and general magnitude of land use and climate change impacts on sediment yield? Is this a consensus opinion? Is this opinion based on specific studies or data sources?

☐ Has the potential impact of extreme storms, including rain-on-snow, hurricanes, or typhoons, been considered in the analysis? Is project design sufficiently robust to manage an extreme flood and its associated sediment and floating debris load?

How sensitive is the project to an error in sediment yield?

Dams for run-of-river projects are designed to manage high sediment loads, including the need to sustain peaking storage while achieving a balance between sediment inflow and outflow. Projects of this type tend to be relatively insensitive to errors in the long-term sediment yield estimate, especially since the year-to-year sediment yield can easily vary by a factor of five. However, in storage projects, or projects that have been designed to depend on costly sediment removal options such as dredging, project feasibility may be very sensitive to errors in the long-term sediment yield estimate.

☐ How sensitive is project viability to possible underestimation of sediment yields?

☐ Will sediment yields 50 percent greater than calculated have a significant impact on project viability?

Sedimentation Patterns and Impacts

What is the anticipated sedimentation pattern in the reservoir?

☐ Has the anticipated sedimentation pattern been analyzed using a calibrated sediment transport model? During a prefeasibility-stage project assessment an empirical method may be acceptable for estimating sediment distribution, but feasibility studies should estimate sediment distribution using a calibrated sediment transport model including sensitivity analysis (see chapter 5).

☐ How robust is the model calibration? (Lack of well-calibrated data sets is frequently a problem at new projects, but at existing dams models can be calibrated against historical sedimentation data including sediment profile, sediment volume, and particle size distribution of sediment deposits.)

☐ Has the modeling been independently reviewed? (An independent review is highly recommended during the feasibility stage of project development.)

What are the important expected sedimentation impacts upstream of the dam?

☐ Has storage loss been projected over time and has it been allocated among the different reservoir pools using the results of empirical analysis or computer simulation?

☐ Have potential long-term upstream sedimentation impacts been evaluated? These include the impacts of delta growth on upstream flooding, waterlogging of upstream soils, navigation impacts, reduced clearance beneath bridges, burial of upstream intakes, and impacts to riparian wetlands and ecosystems. Recall that the delta can extend for several kilometers upstream of the original pool limit. Modeling is required to address this question.

☐ Will operational rules need to be modified in the future to accommodate sedimentation? Has an analysis been performed of the operational changes that will be required and when this may occur?

What are the important expected sedimentation impacts downstream of the dam?

☐ Have simulations been made of long-term downstream channel incision and coarsening of the bed (including armoring) resulting from the upstream trapping of bed material by the dam?

☐ Have impacts of bed incision been evaluated and vulnerable communities or downstream infrastructure and environments been identified? Impacts may include increased bank erosion, scour at bridges and other river infrastructure such as water intakes, and reduced sediment load to the coastal regions leading to increased coastal erosion.

☐ Are mitigating measures anticipated? What will they cost and who will pay for them?

☐ Are significant ecological changes anticipated due to dam-induced modifications to the sediment regime, which may include reduced turbidity, upstream trapping of organic material serving as food supply, change in downstream nutrient releases, and change in river bed material composition downstream of the dam?

Sustainable Sediment Management Measures

☐ Have the sediment management alternatives enumerated in figure 7.1 been considered?

☐ Has a sustainable sediment management plan been developed to identify the management strategies that may be used over time to combat sedimentation?

☐ Which measures will be implemented to enhance sustainability and what is the implementation schedule? Some techniques, such as turbidity current venting, may be used starting in the first year of operation; other techniques may not be appropriate until sedimentation is more advanced.

☐ Are the dam, intakes, and other hydraulic structures designed to facilitate implementation of future sediment control measures?

☐ Is the intake properly designed to address the five performance standards summarized in figure 8.2: passage of all floods including hazard floods; passage of ice and floating debris; passage of sediments; bed control at intake; and exclusion of sediments, floating debris, and air (when necessary)?

☐ Have abrasion-prone civil and hydromechanical elements been designed to facilitate access for repair and replacement?

☐ Has the hydropower cooling system been designed to handle existing and future suspended sediment concentrations?

☐ What is the anticipated cycle for repairing hydromechanical equipment affected by abrasion?

☐ Has the need for a real-time sediment monitoring system and sediment-guided operation been evaluated, and if needed has it been incorporated into the project?

☐ Is there a viable end-of-project scenario? Simply walking away from large dams full of sediment may not be possible, particularly when the river discharges coarse material that will gradually erode and destroy spillways or other components. Sediment loading against the dam may also impair dam safety during seismic events.

☐ Does the project design incorporate resiliency against potential catastrophic events?

☐ Has a reservoir monitoring program been developed that includes a standardized bathymetric protocol starting with the first bathymetric survey soon after initial filling?

☐ Is it important to continue monitoring sediment inflow in the river upstream of the dam, and if so who will pay for this?

☐ Has a monitoring program for impacts downstream of the dam been designed? Who will implement the monitoring? This would, at a minimum, typically include repeated measurement of river cross-sections and documentation of changes in bed material grain size.

Extending the Life of Reservoirs • http://dx.doi.org/10.1596/978-1-4648-0838-8

Development Paradigm

☐ Is the project based on a sustainable use approach or on a design life approach that would lead to eventual abandonment?

☐ If it is based on a design life approach, can the project design be amended to facilitate sustainable use? Which sediment management approaches aimed at prolonging sustainable use of the dam and reservoir can be implemented?

☐ How do climate change and reservoir sedimentation jointly affect project benefits, that is, reliability of water and power supply?

☐ If sustainable use is not considered an economically viable option, what is the decommissioning scenario and how will these costs be borne?

☐ Does the climate change assessment account for future changes in both mean flow and hydrologic variability? How do changes in mean flow and hydrologic variability affect the reliability of water and power supply? How does it relate to the reduction in reservoir volume due to reservoir sedimentation?

☐ Does the economic analysis of the project acknowledge the essence of sustainable development, that is, creation of intergenerational equity?

www.ingramcontent.com/pod-product-compliance
Lightning Source LLC
Chambersburg PA
CBHW080423270326
41929CB00018B/3133